STEEL

STEEL

THE STORY OF PITTSBURGH'S IRON & STEEL INDUSTRY 1852–1902

DALE RICHARD PERELMAN

Published by The History Press
Charleston, SC
www.historypress.net

Cover: *Vertical Pour*, courtesy of Ron Donoughe.

Manufactured in the United States

ISBN 978.0.73850.355.4

Library of Congress Control Number: 2016948307

CONTENTS

PREFACE

The book *Steel* juxtaposes Pittsburgh's wealthy iron and steel manufacturers of the nineteenth century against the immigrant labor force who toiled in their mills and mines. Titans Andrew Carnegie, Henry Oliver and Henry Phipps lived in poverty as children. Most of these industrialists were first- or second-generation Western Europeans—Scottish, English, Welsh, Scotch-Irish or German Presbyterians—and imbued with a strong spirit of capitalism and the belief in their own God-given superiority. In contrast, the laborers in the mills and mines generally were Eastern European—Russian, Polish, Hungarian and Slavic Catholic or Orthodox peasants. The bosses displayed minimal concern for the safety and welfare of the unskilled workforce, who suffered the ravages of disease, inadequate sanitation, poor nutrition and a dangerous workplace.

Although abundant information is available on plant owners like Andrew Carnegie, B.F. Jones, Henry Oliver, Henry Phipps, Charlie Schwab and Henry Clay Frick, relatively little information exists on the worker population, many of whom spoke little or no English. I have relied heavily on the descriptions told to me by their descendants and thus have taken a degree of poetic license with the details of their stories, although the basic truths are factual.

I wish to thank my wife, Michele, for her unwavering support, excellent editorial assistance and photography, as well as my two children, Sean and Robyn, who each are very proficient writers. My appreciation extends to the Yale Writers Conference and to my two instructors, M.G. Lord and Hope

Dillon. I especially enjoyed attending Yale with my good friend and guru-novelist Phil Gasiewicz. I also wish to acknowledge artist Ron Donoughe for allowing me to use his oil painting *Vertical Pour* for my cover. Graphologist Dr. Peter Mancino delivered spot-on personality analysis of the major characters. Reader Cynthia McNickle gave me the encouragement I needed. Editor Heather Lundine helped shape the story, and curator Heather Semple twice led me and my wife on a Duquesne Club tour and permitted the reproduction of Aaron Harry Gorson's 1903 painting *Mills in Winter* and Frank DeAndrea's 1885 painting *The Blast Furnaces of Pittsburgh*.

Finally, I would like to thank Matt Todd and Julie Foster at Arcadia Publishing and The History Press for their work to shape the final manuscript.

CHAPTER 1

IRON

The J. Edgar Thomson steel plant was the finest in Pittsburgh, probably in the world, and its superintendent, Captain William Richard Jones, was the country's most celebrated steel man. He ran an efficient plant, the pride of the Carnegie Empire. The short, tough, fifty-year-old superintendent had worked his way through the ranks, starting as a machinist's mate at age ten. He had enlisted as a private in the Union army during the Civil War and earned a battlefield commission to captain. Now, he managed the steel industry's largest mill.

One Thursday evening, September 26, 1889, six of the plant's seven furnaces were humming. Day and night, J. Edgar Thomson pounded out record output under Jones's firm hand. The second shift had gone smoothly—except that damned Furnace C. It had been acting up all day.

"Cap'n, come quick! Mr. Gayley needs you," yelped a sweaty laborer.

Jones did not tolerate production snags, whether caused by man or machine. Getting into the action was part of the job. "Let me take a look," he grunted. A clump of slag and coke had blocked the furnace like a flotsam and jetsam of logs damming the flow of a river. The team had to react quickly to avoid an explosion from the building pressure.

Sparks flew and the furnace roared as the men tapped the line with a sledgehammer to reduce pressure. "Keep at it," exhorted an impatient Jones.

"Shit, it ain't tapping," shouted a Hungarian laborer, pounding furiously at the metal rod held in place by worker Michael King. "We're in trouble.

Pressure's a buildin' fast!" The struggling workers stepped aside to let Jones take a look.

Jones entered the fray with the same intensity he had demonstrated as an officer with the Eighty-Seventh Pennsylvania Volunteers at Chancellorsville and Fredericksburg. He grabbed the hammer from the Hungarian laborer and swung with all his strength to attack the blockage.

Without warning, a one-foot metal sheet just above Jones's head flew from the furnace, spraying a deadly wave of cinder ash and molten steel. The onrush encased the unfortunate Hungarian laborer in a steel ingot. His body would not be discovered until the following day. Michael King dissolved, boiled in a lava flow of steam, buried in a shipment of rails. The wall of liquid death showered Jones in a bath of molten steel, scalding his face and hands.

Jones leapt for safety, but he struck his skull on a nearby rail car. The team reacted swiftly. A foreman shut down the blast to the furnace, halting the hail of flames and fire—too late. A nearby worker dragged the semicomatose Jones to the safety of an empty room, where a physician treated his burns and summoned a gurney. Jones moaned incoherently. He never regained consciousness and died in the Homeopathic Hospital three days later.

Thousands attended the superintendent's funeral on October 2. The "Cap'n," as the men called him, had been no ordinary manager. He got his hands dirty and knew steel. He would chip in when the going got tough, and that's what killed him. "He could be a bastard, tough, ornery and mean, but at least he was our bastard," spouted one worker, summarizing the general feelings of the plant. "Kelly, one of the two others killed in the accident, had been a good Mick, but now a dead one," eulogized another worker. "There but for the grace of God lay I, on my back in a pine coffin," thought another laborer as he cocooned in the safety of his row house after the funeral.

Carnegie Steel executives Henry Clay Frick and Andrew Carnegie served as honorary pallbearers, although Jones had liked neither man. Workers from the plant escorted the coffin to what is now called Homewood Cemetery.

Only family and close friends would mourn the laborers, Michael King and the nameless Hungarian. Carnegie could easily replace an unskilled worker with another hungry body, but Jones's death proved downright inconvenient for the team at Carnegie Steel. He had been an asset who drove plant production to record levels and developed dozens of valuable patents, including a "Feeding Apparatus for Rolling Mills," an "Apparatus for Removing and Setting Rolls" and a "Hot Bed for Bending Rails." The "Mixer" Jones invented combined several steps in the manufacturing process, saving hundreds of thousands of dollars while providing J. Edgar Thomson with a distinct competitive advantage over its rivals.

The day after the funeral, the pony-sized bundle of nerves Henry Phipps, a Carnegie partner and financial officer, accompanied by Henry Curry, the plant's furnace supervisor, knocked on widow Harriet Jones's door, hat in hand. The executives ostensibly came to offer their condolences, but they had an ulterior mission. Phipps carefully explained how all patents filed on company time belonged to Carnegie Steel. He glanced toward a picture of Captain William Jones in his Civil War uniform on the mantel before shifting his gaze. A tear formed on the widow's cheek, and Phipps felt a twinge of sympathy. Harriet Jones already had lost two of her four children at an early age and now a husband. Phipps continued, "As a reward for your husband's years of service and his unfortunate untimely death, the company graciously is offering a lump-sum settlement of $35,000, provided you sign over all Captain Jones's patents." Harriet Jones knew $35,000 represented a fortune, and possibly a fair price, even though her husband had earned the unheard-of salary of $25,000 per year. The amount offered would feed and clothe her family for as long as she lived. The grieving widow signed the document without objection, anxious to have these men go and grateful for the money.

Henry Phipps raised himself from a chair to his full five-foot, three-inch height, took leave of the widow and departed the house with a complacent smile. Jones was dead, the widow was paid and Phipps and Curry had recovered the invaluable patents for a fraction of their value. Steelmaking at the plant would continue without disruption, and Carnegie Steel's immense profits would continue.

• • • • •

The iron and steel industry dominated the economy of nineteenth-century Pittsburgh. In the last quarter of the era, the J. Edgar Thomson plant under the leadership of Captain Jones dwarfed the competition. While metal production served as the primary impetus for the city's progress and wealth, heavy industry delivered horrible side effects. The noxious stink, searing heat and noise from the mill sucked the common laborer dry. Writer Hamlin Garland described the city of Pittsburgh, home to several of the country's leading iron and steel mills, as like "looking into hell with the lid taken off."[1]

The loud click-clack of horse-drawn wagons against cobblestone, the hammer striking the blacksmith's anvil and the shouts of street vendors hawking wares intermingled with the whinny of horses and the snap of a whip, generating a cacophony of noise. The smells were worse. Scraggly chickens clucked and wild pigs oinked while covering the axle-rutted byways with disease-ridden feces. Pigeons blackened the skies and dropped excrement below them, spreading their own pestilence. A single large horse expelled twenty-four pounds of dung daily, attracting rats, flies and other noxious vermin. A decomposing horse carcass might rot unattended for days on the dirt streets until carted away by the authorities or its owner.

Without sewers and running water, disease from the streets ran rampant. Families tracked contagion into their homes on the bottoms of their shoes. April rains splattered the mess into basements, delivering a lethal dose of germs. Pittsburgh's winter slush and drafty rooms brought pneumonia, flu and bronchitis. Summer carried its own horrors. Local markets hawked rancid squirrel, rotting hog and week-old venison. Spoiled meat and the metallic tang of factory waste permeated the dinner table. A diet low in fruits and vegetables led to an unhealthy lifestyle. Hundreds died from cholera, diphtheria, dysentery, tuberculosis and plague. The scourge of bad food, tainted water and the resultant diseases assailed rich and poor alike. Fanny Jones, a niece of iron baron Ben Franklin Jones and the fiancée of wealthy banker Andrew Mellon, succumbed to tuberculosis in her twenties. Even mighty Andrew Carnegie barely survived a bout with cholera.

Hamlin Garland also described the steel town of Homestead, a typical mill city located a few miles from Pittsburgh: "The streets were horrible; the buildings poor; the sidewalks were sunken and full of holes; and the crossings were formed of sharp-edged stones like rocks in a riverbed. Everywhere the yellow mud of the streets lay kneaded into sticky masses, through which groups of pale, lean men slouched in faded garments, grimy with the soot and dirt of the mills."[2]

Management displayed a callous neglect for safety. The filth and danger at the mills proved even more pernicious than that encountered on the surrounding streets. Scalding water leaked from the ceiling pipes and oozed onto the floors below. Men either learned to dodge the sizzling steam or bore reminders of it for a lifetime. Iron making was dangerous, but the advent of steel would magnify the risk many fold. Ladles of molten metal swung precariously from chains, threatening instant eternity. Crushed limbs and broken bones occurred daily due to tools dropping, slips, falls and malfunctioning equipment. In one single year in the late nineteenth century, 195 industrial accidents in western Pennsylvania culminated in death.

After a long day's work in this treacherous environment, weary Hunkies, Poles, Slavs and Wops, as their Western European bosses pejoratively called them, found solace from their drudgery and danger in saloons. The more vocal, emboldened with a shot or two of whiskey and a beer chaser, might hatch plans for unionization. The common laborer struggled to eke out a living. "We work like dogs and get paid like whores," bellyached a Slav in broken English—a brave militant when far from the earshot of the bosses.

The average immigrant laborer lived in ramshackle housing. Single men crowded in rented rooms or lived with family. Few homes had running water. Wives and mothers dragged drinking and bath water from the Monongahela, the same river where the mills dumped their industrial waste. Families routinely bathed in the chilly Allegheny and Monongahela before city council outlawed the practice. Laborers from iron manufacturer Jones and Laughlin squeezed into South Side areas like Hunky Hollow and Painter's Row, which contained five hundred sardine-sized dwellings, most with no lawn and only a rudimentary cellar kitchen. Other workers lived in Allegheny City, now a part of Pittsburgh and home to Heinz Field and PNC Park. Braddock and Homestead contained their own overcrowded row houses. The location made little difference—slums were slums.

Coal miners in nearby Connellsville, the immigrant grunts who supplied the basic ingredient for the coke used as fuel, fared even worse than mill hands. Those lucky enough to survive the trauma of cave-ins and mine explosions died young from black lung disease, overwork or poverty. Some unscrupulous owners cheated these workers. With wage rates based on coal weight, managers might count 2,100 pounds as a ton. Miners frequently found themselves short of cash. Employee stores provided credit but charged usurious rates for basic necessities like flour and salt. Money-hungry owners forced workers to sign yellow-dog contracts, a pledge not to join unions or face the penalty of an immediate loss of job.

Writer Garland described the dehumanization of mill and mine life: "The worst part is that it brutalizes a man. You start to be a man, but you become more and more a machine. It's like any severe labor, it drags you down mentally and morally just as it does physically."[3] "No wonder we pray to Our Lady of Sorrows," complained one Slovak wife.

As a response to the horrors in the mills, the iron and steelworkers invented the mythical hero Joe Magarac, cousin to Paul Bunyan, the logger hero who rode the giant ox Babe. With his narrow waist, thick chest, calloused hands and muscled biceps, Magarac personified the physical strength required to survive in the steel mill:

Joe Magarac could bend steel rails with his bare hands.

Born out of Braddock, earth, rock and hill, king of the ingots, pride of the mill, nothing about him was timid or small. He gathered the scrap iron, the limestone, the ore; fanned the white heat to an angry red roar. He poured liquid fire in an ingot mold, and taking a handful before it got cold, he squeezed through his fingers and watched it congeal from taffy-like ribbons to straight rails of steel. Best steelmaker in the land, steel-heart Magarac, that's the man.[4]

Like their mythical counterpart Joe Magarac, Pittsburgh's unlettered iron and coal workers had fled their villages and towns in search of a brighter future in America. Ukrainian farmboys avoided conscription and death in the Russian army by hiding beneath wagons of hay headed for port cities like Odessa. Hungarians, Poles, Serbs, Slavs and Croats likewise fled the Old World in search of greater opportunity in America. With a few kopecks in their pockets, they bought third-class passage on tramp steamers or sailing crafts headed to New York. Stuffed like sardines in cargo holds, these brave souls battled scurvy, storms, high waves and disease. Somehow, this hodgepodge of humanity reached the final leg to Pittsburgh.

Few Eastern European immigrants spoke much English, placing them at a distinct disadvantage. As Catholics in a Protestant-dominant industry, they met up with a heavy dose of prejudice in an industry run by their British and German bosses. Heavy accents and limited communication skills doomed most to bare subsistence jobs without promotion—cleaning floors, hauling slag and scrap and loading ore, limestone and coke into the furnaces. Complaints served no useful purpose. Only acceptance and the ability to learn the language of steelmaking might vault the occasional worker toward advancement at higher pay.

These dispirited men trekked six days a week across the bumpy paths between the crowded ghettos they called home and the mill. Long twelve-hour shifts and hazardous work provided the possibility, even the probability, for injuries. Tired men make mistakes, and iron and steelmaking extracted a fierce price. Few who worked in the iron and steel industry survived past sixty. Since unskilled labor earned seven dollars per week or less, those with wives and children faced a life of meager subsistence. A better-paid skilled tradesman might skimp and save to squirrel away enough to buy a small house. The unskilled continued in rented rooms or, even worse, lost their jobs through ill health, injury, union activity, technological displacement or just plain bad luck.

Day after day, the roar of the furnace, the clang of chains, the pounding of the trip hammer slamming against an anvil, the thud of ore pouring into the hopper and the hiss of steam buffeted the ears of the workers. Smoke, heat and the acerbic stink of sulfur assaulted their noses. The glare of the blow during conversion and the haze that followed burned their eyes.

Like the fiery conversion of pig iron into steel through beating and heating, the country bumpkins who entered the mill soon found themselves tempered into hard metal. Survival required quick reflexes, strength and thick skin. Those who entered as boys quickly hardened as men. Numbness, only numbness, made mill life tolerable. At workday's end, the married man might flee the clatter of the mill to seek solace in the quiet warmth of his house, a loving wife and possibly the adulation of an adoring son or daughter. The laborer without family might seek escape by wasting his hard-earned dollars in the seedy bars or brothels surrounding the plant. The pressure from a real or imagined insult often exploded like a Bessemer heat into a fight. It mattered little who won or lost because the next day the combatants would return to the hell of steel for another bout with the devil.

On Sundays, the godly attended Latin Mass at St. Thomas in Braddock and thanked Jesus for the blessings of the past week, a family, a roof over their heads and enough to eat. The most ambitious learned to read English but, more importantly, encouraged their children to go to school. Natural curiosity allowed the lucky few to advance to positions like senior smelter or furnace attendant, but these proved the exception. All prayed for a better life for their children.

In juxtaposition to the workforce, the bosses lived a privileged life, their bellies filled with beef, potatoes and fresh milk. Ketchup, mustard, pickled onions and sauerkraut, all made in Pittsburgh, most courtesy of local purveyor Henry Heinz, added spice to their tables. Decked out in top hats and fancy suits, the steel barons paraded about Pittsburgh.

The bosses were different than the working folk—not just richer. They weren't necessarily evil men, just men who lacked sentiment and pursued the dollar regardless of the consequences. They considered labor a mere tool, a shovel to hoist cinder ash, to be discarded when the blade cracked or the handle broke, thrown aside like useless slag.

The early iron and steel bosses such as Ben Franklin Jones, Andrew Carnegie, Henry Oliver and Henry Phipps subscribed to the theory of manifest destiny, that God ordained them with what philosopher Max Weber called the "Protestant Ethic." Material accumulation in this life pointed to a future place in heaven. American capitalism developed hand in hand with John Calvin's teaching of predestination. Metal-monger B.F. Jones assumed the Almighty not only justified his pursuit of wealth but pronounced it a virtual commandment. The hand of the divine touched those who prospered. When asked how he obtained his wealth, Cleveland multimillionaire John D. Rockefeller proudly proclaimed, "God gave me my money."[5] Popular songs crooned the tune of material achievement. The nineteenth-century rich mouthed the words of the song "Dad's a Millionaire": "Good-bye to poverty, want and care. The fortune's come. We've waited so long, and Dad's a Millionaire."[6]

The majority of the Pittsburgh steel and iron barons were Scotch-Irish, highly focused on the pursuit of business. These Protestant entrepreneurs indulged in few diversions. They ruled their homes and businesses with an iron rod, counting pennies and expecting others to follow their example. Titan Andrew Carnegie bragged, "America would have been a poor show had it not been for the Scots."[7]

• • • • •

Ben Franklin Jones (no relation to Captain William Jones) led the charge as a pioneer of Pittsburgh's mid-nineteenth-century iron industry. The ninth child of Jacob and Elizabeth Goshorn Jones, B.F. was born on August 8, 1824, in Claysville, Pennsylvania, forty miles south of Pittsburgh. His Welsh father, Jacob Jones, named him for Revolutionary hero Benjamin Franklin.

Jacob Jones had spent his youth meandering from western Pennsylvania town to town scavenging for work—surveying, farming and running a small rooming house—whatever it took to feed his family. From Claysville, the peripatetic Jones family eventually settled in New Brighton, north of Pittsburgh. There fourteen-year-old B.F. attended the New Brighton Academy, where teachers drilled basic reading, writing and arithmetic through his head, mostly through the *McGuffey Readers*, which reinforced the strict Christian virtues of thrift and philanthropy. His mother's biblical parables rounded out his ethical instruction.

After school, Ben helped at the family-owned rooming house, where he learned the value of a pleasant "Good morning, sir" and "Have a fine day, sir." His surveyor father, Jacob Jones, instructed him on the critical necessity of accuracy, demonstrating how he marked the parcels he surveyed with iron rods to delineate property lines. The father always etched his initials for identification. Jacob Jones further cautioned his son: "Tiny errors might create disastrous property disputes. Mark your own work with pride just as I do." B.F. promised to strive for the first tier.

Ecclesiastes in the Good Book proclaimed that all things have a time and a place. Primed with Protestant morality and an ambition to succeed,

Ben Franklin Jones.

the eighteen-year-old left home for Allegheny County. This would be Jones's time and Pittsburgh his place.

B.F. found himself sitting on a dock in 1843 along the Monongahela River, where he inhaled the surrounding sights and smells. Serpentine smoke rings slithered across the sky as a worker burned the guts out of a log to construct a dugout boat while humming a song. Beefy carpenters hacked at wood logs for primitive single-trip, flat-bottomed barges to carry coal to New Orleans. The staccato beat of a hammer pounding out a glassware container meshed with the sonorous tones from a distant fiddler to improvise an industrial symphony. As Jones absorbed the richness of the scenery, he recognized the importance of water transportation to nineteenth-century Pittsburgh. Roads outside the city were primitive, limiting wagon cartage. Pennsylvania had no railroads yet.

In addition to the Monongahela, Allegheny and Ohio Rivers, horses and mules towed cargo barges over the thirty-foot-wide by four-foot-deep, decade-old Pennsylvania Canal system. Bakewell, Pears & Company glassware from Pittsburgh found its way to New York or Philadelphia on a barge. Fabric from New York or saws from Philadelphia flowed to Pittsburgh by water.

Jones spotted a sign for Kier, Royer & Company, a leading operator of the Mechanic's Line of coal-transport canalboats between Philadelphia and Pittsburgh. Summoning a dose of courage, he rapped on the door and approached Samuel Kier for a job. Kier, impressed by the boy's get-up-and-go, offered B.F. room and board, an unpaid apprentice's position. Jones, in turn, promised Mr. Kier he would never regret his decision.

Jones showed up early and stayed late. He tended to details and learned from his mistakes. When Kier witnessed his protégé calm a shrieking customer over a late shipment, he complimented his apprentice: "Son, you've got a firm head on those shoulders. Keep up the good work. You'll go far."

B.F.'s effort paid dividends. He treated the business as if it were his own, preserving the bottom line like a mother hawk defending her chicks. Although he received no wages for the first year, he earned an invaluable education in practical shipping management. By age twenty, he had changed from a boy to a man, reaching five feet, ten inches in height. Customers liked and trusted this big bear of a man, who recently had sprouted his trademark beard. Jones's broad face, aquiline nose and hazel eyes suggested seriousness. His acceptance of responsibility generated trust. Samuel Kier had groomed B.F. Jones as his ideal right-hand man.

Mid-nineteenth-century canal trafficking between Philadelphia and Pittsburgh was an arduous and time-consuming trek that took several weeks and crossed a conglomeration of rivers, canals, bridges, tunnels and railways. Stationary steam engines, mules, horses and human power dragged twin-sectional breakaway barges across mountains and down inclined planes.

Although Kier had promoted Jones as a junior partner in the Independent Line, the rapid advance of the Industrial Age brought with it vast changes. The state granted a charter to the Pennsylvania Railroad on April 13, 1846. The first train reduced travel time for the three-hundred-mile canal trip from Philadelphia to Pittsburgh from weeks to just fifteen hours. Rail transport quickly turned the canal system into an obsolete dinosaur, forcing Kier and Jones to seek new careers.

Kier opted for rock oil, a byproduct of his father's salt mine. He first bottled and sold oil as a medicinal panacea before learning how to convert it into kerosene. In 1854, he developed the country's first refinery, earning him a tidy living: $40,000 per year in revenue, or nearly $320,000 in current dollars. Historian Karen McInnis aptly dubbed Samuel Martin Kier the "Grandfather of American Oil."

B.F. pursued a different path: the iron trade. While crisscrossing the state on sales calls for Kier, he had visited numerous iron-ore sites, coal mines, smelters and foundries—all canal transport users. Locomotives, axles and rails required iron. The growing United States population hungered for iron farm implements, tools, cooking utensils, ovens, horseshoes and nails. As early as 1826, Colonel James Anderson, Sylvanus Lothrop and Henry Blake had founded the highly successful Juniata Rolling Mill in Allegheny City, one of western Pennsylvania's earliest big-time manufacturers of nails and semi-finished rectangular beams called blooms. Already, some referred to Pittsburgh as the Birmingham of the United States. B.F. Jones understood the opportunity provided by the burgeoning Iron Age and seized it by acquiring the Buena Vista Iron Furnace and Forge of Armagh, fifty miles east of Pittsburgh.

Although the forge had foundered under an ocean of cheap English iron, weak management and a massive business slowdown caused by the great Pittsburgh fire of 1845, Jones convinced Samuel Kier of the potential for profits with proper leadership. Jones's managerial ability, Kier's political and financial clout and an improving economy provided the perfect formula for a winning business. When Kier's former oil partner, Secretary of State and later president James Buchanan, successfully lobbied for increased iron tariffs, the outlook for the venture appeared even more positive.

Jones eased into the iron industry like an English lord fitting into a bespoke suit. During his evenings, he poured through metallurgical books and articles to improve his theoretical background, while studying his puddlers, the men who stirred the molten iron with metal rods, to absorb practical iron-making knowledge. He tolerated neither guff nor excuses from his workforce but maintained the same high standard for himself.

The Pittsburgh social scene readily embraced the up-and-coming Jones as an eligible bachelor, introducing him to the city's proper single ladies. With a taste of success and love in his heart, B.F. wedded handsome Mary McMasters on May 21, 1850. Mary delivered a host of key family connections as part of her dowry. Her brother-in-law was General William Larimer, a rail-line magnate; a founder of the city of Denver, Colorado; and a future Kansas senator. Mary's niece one day would wed James Ross Mellon, the eldest brother of banker Andrew Mellon. The couple would produce four children: Mary, Elizabeth, Alice and Benjamin Franklin Jr.

As finances at the Buena Vista foundry turned the corner, Jones's confidence magnified. He discussed a merger with German iron makers John and Bernard Lauth, two brothers who owned a modern foundry located near the old Independent Line headquarters. The idea made sense. "We could triple our business by merger," Jones told Kier, who could feel the synergy expounded by his partner. Jones and Kier anted $12,700 in cash, mostly supplied by the latter. The German-born Lauth brothers chipped in puddling furnaces, mills and land along with a healthy dose of iron-making savvy. In a December 3, 1853 agreement, each partner received a 25 percent ownership share. Jones and the Lauth brothers would earn annual salaries of $1,500. Kier, as a silent partner without day-to-day responsibilities, received no salary. The four entrepreneurs slated all profits for reinvestment.

The American Iron Works, as the new firm called itself, sat on one and a half acres along the Monongahela River between two taverns off present-day Carson Street. The plant contained all the necessary basic production equipment: four puddling and two heating furnaces, a guide mill, assorted muck rolls and a crocodile squeezer, in addition to a variety of trade tools.

Ben Franklin Jones eagerly had awaited the formal opening of the American Iron Works. Unfortunately, the funeral of John and Barnard Lauth's father postponed and delivered a somber tone to the festivities. Jones rose early the morning of the grand opening. Excitement surged through his veins. "Mary, this is a great day for us," he told his wife at the breakfast table. Since dress and deportment set the table for a proper gentleman, Jones donned his newest black suit, starched white collar and dark tie ornamented with a trendy pin for the occasion.

This proud rooster reined in his horse and rig in front of the new facility, pausing to take in the moment. He hitched the horse to a rail and surveyed the tiny sign sporting the company name. As he touched the sign, he envisioned a heady future. The plant's puddling and heating furnaces held the key to unlocking his membership in Pittsburgh's iron aristocracy, and B.F. Jones intended to enjoy life's dividends.

The American Iron Works would become Pittsburgh's first combined roller of wrought iron and manufacturer of pig iron, so named for the molten metal flowing through sand castings, which resembled a mother pig suckling her brood. Although other companies in the city performed one of these functions, none did both. Jones had allied himself with two of the most skilled ironmongers in the city and had a brilliant entrepreneur in his third partner, Samuel Kier. The manufacturing of metal demanded physical strength and technical knowledge. Bernard and Ben Lauth excelled at both, and Jones knew how to organize the business.

In the months ahead, B.F. surveyed the mill from his office when he was not on the road selling. Jones, Lauth and Company's production capabilities amazed him. From a distance, he watched the workers scurry in all directions like tiny ants. On closer inspection, distinct forms and functions took shape. A stream of smoke from the refractory furnace snaked its way up the giant chimney. A black-bearded, sweaty laborer hefted a shovel brimming with coke into the brick-lined rectangular hearth lined with cinder ash. As the worker bent over, the blistering heat singed his hairy chin and blackened his face, the flames silhouetting his bare chest and muscular arms. The man coughed and spat a mouthful of phlegm onto the nearby pile of fuel where it landed. A smoldering flame sizzled among the ashes from the spittle. He stepped aside and gasped for a breath of fresh air. It was only nine o'clock. The day had just started in earnest.

A second worker shoveled pig iron into a furnace loaded with limestone and coke. As the six-hundred-pound rabble boiled and chortled to a liquid, another automaton stirred the mixture with a tempered iron paddle to drive off impurities. As the iron cooled and congealed to the consistency of paste, Davy Morgan, a powerful, skilled puddler, wearing a skullcap like a cardinal signifying his exalted status, passed an iron bar through the front of the furnace and hefted the pasty metal sphere, called a "heat," as it became ready for removal. The veins on Morgan's arms pulsated as he hoisted the one-hundred-pound glob of muck iron, shouting, "Wedge out," and handed it off to his young apprentice. When the molten mass cooled, the squeezers and rollers took over, crushing, reheating, hammering, rolling and kneading the metal orb to remove impurities. A man named Joe Manky operated the muck roller, and "Blood Pudding" Frank Holy served as the catcher. Holy snatched the hot iron as it passed through the rollers, ready for a second pass. The violent process of heating and beating produced cast-iron metal ingots weighing up to one hundred pounds, whose carbon content had been reduced through the process from as much as 7 percent to 3 percent or less to improve structural integrity. The cast iron might be used to fabricate Franklin stoves, cooking pots and grids. Further treatment resulted in wrought iron, sometimes known as soft steel, which could be reheated and shaped into nails, tools and pressed or cut spikes.

Puddler Morgan earned a premium wage based on his per-ton productivity. He had learned the secret art and science of iron making from his Welsh family kin. He knew the exact moment to extract the boiling brew, and he guarded his knowledge carefully.

B.F. respected his iron makers, especially the puddlers. The plant was a noisy place, in fact a hazardous hellhole, but the clang of iron against iron brought profit to his pockets and a smile to his face. With danger accompanying each operation, iron making was not for the squeamish. Laborers got hurt and died. Jones understood and respected the courage they displayed on a daily basis. They were brave and strong, but he also believed his own contributions far outweighed those of his crew. He organized the business, paid for the raw materials, marketed the product, collected the moneys and, most importantly, created the jobs. If the puddlers, squeezers, heaters, muck rollers, drag outs, lifters and sweepers played the role of the orchestra, he was their conductor. The men might make the iron, but he directed the show. Jones paid them just enough wages for food, clothing and housing. He believed surplus pay would only

A puddler at work.

culminate in dollars wasted on sin and a curtailment of the company's growth, creating higher industrial unemployment.

Jones, Lauth and Company workers at the American Iron Works in the mid-nineteenth century generally were Irish, Welsh, Scottish and German. This hardy lot attended Sunday morning Mass at the German Catholic Church or at St. Johns, where stiff-willed Father Reynolds presided over the flock in a no-nonsense fashion. After church, the men gathered near the steps of Grierson's grocery to engage in political chitchat. On Friday nights, many soothed their woes at Dominick Maguire's Tavern on Carson Street, where a glass of beer and a whiskey or two loosened tongues and allowed the vocal few to harp about hard working conditions, poor pay, shrewish wives and the weight of responsibility on their backs. They had no union to protect them and small hope for a brighter future.

Jones wasted little time with the complaints of his workforce. He handled sales, distribution and banking for the company. He spent much of his time traversing the tri-state area of Ohio, West Virginia and Pennsylvania on horseback, coach or rail to move inventory, generally exchanging invoices for trade notes rather than receiving cash or gold in hand. He would discount the notes for working capital at a bank, shipping the ordered iron product to his customers via rail, water or Conestoga wagons. Along the way, he gained the admiration and respect of the banking community. The rising entrepreneur subscribed for shares and received a board seat with Pittsburgh Trust and Savings, a recently incorporated bank founded by wealthy Irish meat purveyor James Laughlin.

Jones recognized that an infusion of cash could be used to purchase new technology and elevate the American Iron Works to the top tier of excellence. Consequently, he painted an intriguing growth picture for banker Laughlin. Without the benefit of dividends or salary, Kier had opted to exit in 1855, swapping his stock for an iron commission business he and B.F. jointly owned. Shortly thereafter, John Lauth retired, and Laughlin stepped in with a $48,000 investment in exchange for a 13/32 ownership interest, identical to that of Jones. Bernard Lauth held the remaining 6/32 minority share. With Laughlin's cash influx, the company, renamed Jones and Laughlin, constructed a two-story office, warehouse and distribution center on Water Street, becoming Pittsburgh's first iron firm with such a high level of vertical integration.

Seeking a tangible symbol to promote the company's superior products, Jones hired English mason James Heakley to top the eight-fluted smokestack of the American Iron Works with an ornamental cover. The Englishman sculpted a man's head topped with a brick crown, visible for blocks. He felt certain Jones would approve. When Jones spotted the crown from his rig, his face darkened and his deep-set eyes glared. He snapped his whip, pushing his horse forward. "Heakley," he croaked. "This is the American Iron Works and in America there are no crowns."[8] The following day, the chastised mason filled in the crown with brick and mortar, altering the smokestack to resemble an American top hat.

Jones and Laughlin's thirty-one puddling furnaces poured out 6,000 tons of iron annually. To ensure the continuity of pig iron and reduce potential bottlenecks, Jones ordered the construction of two giant Eliza blast furnaces, named in honor of James Laughlin's daughter and his own mother, sister and daughter, all of whom shared the same name. The chimney on each furnace soared forty-five feet in the sky with a massive girth of twelve feet. These monster furnaces became the first in Pittsburgh to use Lake Superior iron ore. A load, called a "burden," of up to 3,000 tons of dirt, coke, iron ore and limestone, when heated to 1,000 degrees Centigrade, yielded approximately 1,000 tons of pig iron and 750 tons of waste or slag. The remaining tonnage disappeared up the chimney in the process.

A minor accident led to a huge uptick in the company's fortunes. Barnard Lauth dropped his tongs into a roller, crushing them. At first, the German cursed his bad luck. However, when the metal cooled, it took on a fascinating bright shine. More importantly, tests demonstrated that the iron strength increased by 75 percent. After a few weeks of tinkering, Lauth developed the revolutionary cold-rolled process. In short order, the ironmonger patented a process in 1860 for rolling in two directions, saving "half the time and half the effort," elevating Jones and Laughlin a notch above the competition.[9] With strong management, low pay rates and high productivity, profitability surged.

U.S. Civil War hostilities erupted on April 14, 1861, pumping up the demand for iron. Pittsburgh's 1,191 factories and 20,500 laborers solidified their support of the war effort. Bernard Lauth had retired from Jones and Laughlin. B.F.'s brothers George and Thomas had taken up much of the slack. Without neglecting his responsibilities at the American Iron Works, Jones expanded his community role. He accepted a seat on the board at the Citizens Fire Insurance Company and served on the fundraising committee for the First Presbyterian Church. Although initially backing the Republican Party due to its protectionist stance against iron imports, both he and James Laughlin developed into staunch abolitionists and Northern patriots. Activist Jones volunteered for the Committee of Public Safety under the chairmanship of Andrew Carnegie's eighty-two-year-old neighbor Judge William Wilkins, a former United States secretary of war.

None of his workers considered Jones a soft touch, but when he came upon a troop of hungry enlistees without rations sitting beside the Pennsylvania Railroad tracks, he bought them crackers and apples from the nearby grocery store. He knew the recruits required a steady supply of food. In response, B.F. volunteered to co-chair the Pittsburgh Subsistence Committee along with fellow industrialists William Thaw and Thomas Howe. The trio headed up the effort to feed and clothe troops prior to formal mustering. They set up a dining hall at the Leech Warehouse on Penn Avenue that served nearly 400,000 meals during the war years.

Jones invested his soul into the war effort. He contacted congressmen and wrote articles of support for the issuance of legal-tender treasury notes backed by government bonds. He joined the executive board for the Pittsburgh Protective Committee, established to defend against attack. When the government called for fifteen companies of regulars, or approximately six thousand recruits, from Allegheny County, the committee proposed a $50 bonus to speed enlistments. Jones accepted a seat on the fundraising subcommittee and personally inked a check for

$3,000 to equip one entire company, hoping to recoup a portion of his outlay from businesses and friends.

James Laughlin's son George, a senior at Washington and Jefferson College, enlisted in the Union army along with a handful of patriotic Jones and Laughlin laborers. At war's end, Major Laughlin would witness General Robert E. Lee's surrender to Union general Ulysses S. Grant at Appomattox and return home unscathed. Other Jones and Laughlin employees proved less lucky. Straightener Dennis Healey died at Malvern Hill, and laborer Tony Heilig lost his life during the war as well.

On September 17, 1862, the day of the Battle of Antietam, where General Lee's forces halted McClellan's forward march through Maryland at a cost of 22,817 American lives, disaster struck Pittsburgh. Jones sat at his desk reviewing bills in the mid-afternoon. A sudden blast rocked the room. Within minutes, a clerk burst through the door.

"Mr. Jones, the Allegheny Arsenal exploded." While turmoil raced through Jones and Laughlin, Jones attempted to calm his crew. He later learned that seventy-seven workers, mostly women who rolled ammunition cartridges for $0.50 to $1.10 per day, had died in the fiery inferno, many burned beyond recognition. This became the city's worst industrial accident ever and shook the citizenry to the core. Rumors also spread of a possible attack on the area's iron, shipbuilding and railroad centers by the Confederate cavalry under General J.E.B. Stuart. B.F. Jones, as a member of the Protective Committee, immediately issued plans for strengthened defensive measures. All the while, Jones pushed his company and the entire iron industry to up the production of armaments and supplies to support the war effort.

Jones and Laughlin output reached record levels during the war. Labor watched the big dogs like Jones lap up Civil War profits. The skilled craft workers hungered for their share of the feed. "They gets their dollars, and all we want is fairness," groused one militant puddler, whose face reddened as he spoke. The boilers and puddlers at the American Iron Works banded together to fight for better wages and working conditions, organizing the Sons of Vulcan Union and electing Miles Humphreys as president. With business booming, Jones opted to recognize the union rather than risk a plant shutdown.

• • • • •

Other ironmongers profited during the Civil War along with Jones. In 1858, Prussian-born Anthony Kloman and his younger brother Andrew had opened a small forge on the outskirts of Pittsburgh. The rough-and-tumble brothers specialized in railroad axles made from scrap iron and earned a reputation for personal toughness and flawless workmanship. Cash poor and facing a backlog of orders, the brothers required a second trip hammer to pound out more axles to meet demand. Unable to procure a conventional bank loan, Anthony and Andrew turned to Tom Miller, the purchasing agent for the Pittsburgh, Fort Wayne and Chicago Railroad. Miller agreed to provide $1,600 in exchange for a one-third interest in the company.

A handshake sealed the deal, but Miller faced a problem. His ownership might create a possible conflict of interest with his employer, one of Kloman's largest customers. To resolve the issue, Miller asked boyhood friend Andy Carnegie to hold the stock in his name, which Carnegie agreed to do. The Kloman brothers vetoed the arrangement due to Carnegie's wheeler-dealer reputation. This decision set in motion a chain of events that would change the lives of Henry Phipps, Andrew Carnegie and the entire iron industry.

Carnegie suggested twenty-year-old former neighbor Henry Phipps as a substitute. Since the Kloman brothers needed a bookkeeper, they readily accepted the innocuous-looking youth, a born analyst of minutiae who viewed business through a microscope rather than a magnifying glass. Phipps possessed another crucial gift—the luck to be in the right place at exactly the right time.

Born on September 27, 1839, Henry Phipps became friends with his neighbors Tom Carnegie, Tom Pitcairn and Henry Oliver, growing up as kids in Allegheny City's poorest section. His older brother John attended school with Andrew Carnegie and Tom Miller. Phipps possessed a serious nature, inheriting his attention to detail from his mother and his work ethic from his English shoemaker father. His parents inculcated him with the Protestant dogma: life without labor lacks meaning. Only long hours and a strong focus could free him from poverty and lead him to salvation. He quit school after the seventh grade and hawked newspapers, delivered shoes, worked at the cobbler's bench and ran errands for a small jewelry store, where he encountered his first business crisis. A customer passed a bogus ten-dollar bill, and his boss most certainly would demand the replacement of the loss—two full months of wages. Panic seized Phipps. He resolved to act quickly. A county fair had filled the area with strangers. On a hunch, Harry dashed from the shop toward the fairgrounds in pursuit of the miscreant who had cheated him. He cornered his quarry and reclaimed the merchandise in exchange for the counterfeit currency.

To advance his employment and pay, Phipps borrowed twenty-five cents from his older brother John and placed an advertisement in the May 10, 1857 issue of the *Pittsburgh Dispatch*: "Willing lad wants work." The ad brought an offer to be an office boy at Dilworth and Bidwell, a purveyor of iron railroad spikes and DuPont powder. He drafted a note to future Pittsburgh Plate Glass entrepreneur and friend John Pitcairn describing his job: "I am at an office on Water Street as a clerk. I make out bills and take orders for powder. I am learning bookkeeping and like my place very well."[10] He studied double-entry accounting in night school with Allegheny City cohorts Henry Oliver and Thomas Carnegie to improve his skills.

Frugality became second nature to the cash-poor Phipps. Long hours and low wages forced him to hoard pennies. He walked to work six days a week rather than pay a few cents for transportation, trekking from Allegheny City across the St. Clair Bridge, oblivious to smoke, cold or heat. When dining out, he ate at a restaurant serving bread with its codfish balls, providing adequate nutrition and the opportunity to sneak an extra slice into his coat pocket. "Save a nickel; save a dime. Thrift is my ticket to prosperity—my pathway to a better life," Phipps chanted to himself. He had watched his father and mother struggle to put food on the table. He hated poverty, the ugly specter of uncertainty and need. He had translated penny pinching to his business persona as well, making parsimony an asset. He diligently safeguarded the smallest details at Dillworth and Bidwell, earning himself the prize of a minority partnership in the powder business before his twentieth birthday.

His brother John, who died in an unfortunate horse-riding accident at age eighteen, had offered his younger sibling a prescription for future success before his unfortunate passing: "Life was not given to be frittered away in dreamy indolence. There is room enough in this great world for all of us to exercise our talents and energy. So much is there to be learnt, so much work to perform, that I think the span of existence is scarcely long enough to accomplish it."[11] Henry's brother William became a Methodist minister. "He quit the bench [cobbling shoes] for the pulpit and rather than working on soles was working on souls," the family joked.[12] Sister Amelia would marry John Walker, a future steel and coal executive. Henry envisioned an equally cosmic future for himself as well.

Many underestimated this stoop-shouldered, pale, plain man who stood a few inches above five feet in height. That could be a grave error. His cold, gray eyes and razor-like stare projected his true nature. Phipps discovered an unlikely advantage in small size. The equally tiny Andrew Carnegie, who described Henry as a "bright, clever lad," gravitated toward undersized men like iron to a magnet.[13]

Tom Miller liked Henry Phipps and offered him half his stake in the Kloman Company, or a 16.5 percent interest, for $800. Phipps appreciated the gesture but lacked even $100. His shoemaker father offered to mortgage his home as collateral but developed cold feet. Miller stepped up and loaned Phipps the money with the agreement that he would repay him from future profits.

Success came quickly to the newly expanded company. The Fort Wayne Railroad's axle purchases plus sales to non-competing firms like Toledo's Whittaker and Phillips, Chicago's Jessup Kennedy, Detroit's Haskell and Barker and Dayton's Barney Parker boosted sales and profits. Phipps tended the books while Andrew and Anthony Kloman pushed out output from the forge.

The ongoing Civil War drove up demand for movable cannons and supply wagons as well as locomotive axles. Metal shortages swept the North, pushing prices from two to twelve cents per pound. A need for additional production capacity drove the Kloman brothers to find a larger plant. The new facility accommodated four puddling and four heating furnaces; three boilers; a large steam engine; four smaller engines; a steam, a trip and a tilt hammer; a train of bar rolls; a set of muck rolls; a squeezer; three blacksmith forges; four lathes; a drilling machine; a screw cutter; shafting; pulleys; and miscellaneous belting, in addition to a full complement of tools. The entire property was capitalized at $80,000. Phipps had become a businessman of consequence, and he intended to maintain his momentum.

• • • • •

Any discussion of nineteenth-century iron and steel must include Andrew Carnegie, who would become the largest stockholder of the country's biggest steel producer. Although small in stature, he demanded respect. He believed the universe orbited around his wants and needs, awaiting his beck and call. He could be sarcastic and volatile, and his nitpicking often irritated those around him. Many admired this sultan of steel. Fewer liked him. As one example, following an unsolicited December 1912 lecture from Andrew Carnegie on how he might have improved his administration, President William Howard Taft jotted on the margin of the letter, "Isn't it pleasant to be told how it could have been done." He forwarded the note to Secretary of State Philander Knox, who wrote back, "As an

Andrew Carnegie.

exhibition of ignorance, mendacity and impudence, this communication of Mr. Carnegie's is quite up to his well known and well deserved international reputation for these mental and moral failings."[14]

Andrew Carnegie was born on November 25, 1835 in Dunfermline, Scotland, the first son of Margaret Morrison and Will Carnegie. Weaned on a hodgepodge of his father's and grandfather's radical working-class philosophy, the poetry of Robert Burns and his Uncle Lauder's romanticized stories about Scottish heroes William Wallace and Robert the Bruce, little Andy developed a high degree of entitlement. Nicknamed "Naig," short for Carnegie, he believed the blood of noblemen and kings surged through his veins.

Andy's father recalled the day he carried his six-year-old son on his shoulders up a hill across the Scottish highlands. "Andy, my boy, I must rest." The father set his son down and asked, "Can you walk a bit?" The father had toted his son nearly three miles. The boy replied without hesitation, "Ah, father, never you mind. Patience and persistence make the man." The easygoing Will Carnegie laughed and hoisted the lad back on his shoulders. Throughout life, Andy would employ others to carry the load.[15]

Work dried up in Scotland for wee, fair-haired political activist Will Carnegie. Demand for his hand-woven damask tablecloths had disappeared with the advent of the Jacquard power loom. "I can nae more get work," he groused.[16] Facing a life of poverty in Scotland, Will Carnegie borrowed money from family and friends for steerage tickets on the eight-hundred-ton triple-masted schooner the *Wiscasset*, bound for America. The voyage from Glasgow proved arduous for most. Theft, seasickness, disease and cramped quarters bedeviled passengers and crew alike. One poor soul died at sea, but the curious thirteen-year-old Andy and his brother Tom savored the hardship as a shipboard adventure—an opportunity to examine every inch of the vessel. After a fifty-day voyage, the ship landed safely at New York's Castle Harbor on July 15, 1848.

After another month's travel by steamer, canalboat and overland trek, the weary expatriates reached Allegheny City, where Margaret's sisters lived. Andy's aunties, Kitty Hogan and Annie Aiken, welcomed their impoverished kin, providing food and shelter. Will eked out a few dollars weaving tablecloths on his brother-in-law's abandoned loom. Margaret, the daughter of a shoemaker, bound shoes at four dollars per week for neighbor Henry Phipps. Her meager earnings somehow kept the family afloat.

Andrew Carnegie's mother, the raven-haired, black-eyed "wiselick" Margaret, who stood half a head taller than her husband, ran the house with a firm hand. Her belly burned with ambition, and she hated the slummy housing she shared with her sister at 336 Rebecca Street in Allegheny City. She egged on her boys by thought and deed. While her husband, Will, dreamed and philosophized, she pushed son Andy to put food on the table. Andy and his younger brother Tom jokingly nicknamed the area "Barefoot Square in Slabtown," but Margaret Carnegie saw no humor in her misery.

When Andy found Ma sobbing over failed family finances, he patted her shoulder: "Ma, someday I'll be rich, and we'll ride in a coach driven by four horses." Margaret slowed her tears and looked at her son. "That will do no good over here if no one in Dunfermline can see us."[17] Andy promised to make his fortune and return his mother home in style. Thirteen-year-old Andy began by seeking a job.

Carnegie set high goals, as did most of his friends. He founded a debating society, named "the Six," that counted Tom Miller, John Phipps, Robert Pitcairn and Will Cowley among its membership. With Andy in charge, the boys argued current events, politics and philosophy. Younger brother Tom palled around with an equally ambitious crew, including Henry Phipps, Henry Oliver and John Pitcairn. Although Andy lacked formal schooling past his thirteenth year, a keen mind and a constant diet of books propelled his self-improvement.

Andrew Carnegie had reached the United States at an auspicious time. America was inching into the Industrial Age in the late 1840s by fits and starts. Pennsylvania had just authorized a railroad system through Pittsburgh. Steam engines had revolutionized the ship industry. Boat manufacturers, glass plants, iron foundries, tanneries and food processors lined the city's three rivers. The recently opened Mercy Hospital provided the latest in healthcare. Rows of telegraph lines connected Pittsburgh to New York, Philadelphia and Cleveland. Although poverty hampered most Slabtown immigrants, a select few possessed the vision, focus and commitment required to take advantage of the opportunities before them.

Starting on the lowest rung, Andy Carnegie labored through six twelve-hour days each week as a bobbin boy at the Blackrock Cotton Mill, replacing full spools of yarn with empties while earning $1.20 per week. Within the year, he resigned for a $2.00-per-week spot at John Hay's mill, a block from his home. The new job included the unctuous task of dipping bobbins in rancid oil. The odor caused Andy to gag and retch, sometimes even vomiting, but he stuck with his work. He intended to escape the hellhole of Slabtown, and nothing would get in his way. The Almighty had predetermined a better life for him.

Carnegie's legible penmanship and ability to handle numbers led to a promotion posting bills in the ledger. To advance his skill, Andy enrolled in a formal accounting course after work taught by a Mr. Williams along with "the Six" friends Will Cowley, Tom Phipps and Tom Miller. Although Andy's career had progressed, he wanted more. Over a friendly game of checkers, Thomas Hogan recommended his ambitious nephew for a delivery boy's job at O'Reilly Telegraph to the firm's Scottish manager, David Brooks. "The lad is special. Ye shall see," he promised.

Carnegie took to his new job like a bear to honey, delivering telegrams to the rich and famous of Pittsburgh. He later wrote in his *Autobiography*: "There was scarcely a minute in which I could not learn something. I felt my foot was on the ladder and that I was bound to climb."[18] When the esteemed B.F. Jones took an envelope from his hand, he patted the boy gently on the head and smiled: "Here you go, Lad." Carnegie stared into his open palm. Jones had tipped twenty-five cents, nearly a third of a day's wages. Andy determined to scout out any messages for his new hero thereafter.

Hard work and good luck impelled Carnegie's fortunes onward. When he found a $500 check, he returned it. The November 2, 1849 issue of the *Pittsburgh Daily Gazette* announced, "Like an honest little fellow, he promptly made known the fact and deposited the paper in good hands."[19] The gesture earned him a cash reward along with a commendation from his employer. With telegraphy on the grow, Andy took the opportunity to recommend several of his Slabtown pals for delivery boy jobs at O'Reilly, including Robert Pitcairn, who one day would advance to superintendent of the Pennsylvania Railroad; David McCargo, a future top executive for the Allegheny Valley Railroad; and Henry Oliver, whose iron ore success would rival that of John D. Rockefeller. Although tiny and somewhat prissy, never growing much taller than five feet, three inches, Andy proved a natural leader. To avoid arguments over money, he convinced his fellow delivery boys to pool their tips. With ambition surging through every pore of his body, he enrolled in a second accounting course to improve his lot. Readings from Shakespeare's plays, Lamb's essays and Bancroft's *History of the United States*, all borrowed from Colonel James Anderson's private library, which he opened for working boys every Saturday afternoon, brightened Andy's weekends and evenings. Someday, he hoped to amass enough wealth to have his own library and loan books to those who could not afford them.

On December 10, 1852, Andy, brother Tom and neighbor Henry Phipps gathered among the crowd of thousands to witness the Pennsylvania Railroad's first train chug into downtown Pittsburgh from Philadelphia. The world was changing rapidly, and Andy Carnegie intended to take part in the revolution.

Andy had impressed his superiors, who advanced him from delivery boy to telegrapher at four dollars per week. He wrote, "Whatever I engage in, I must push inordinately."[20] As the family's chief breadwinner, he aimed high. Once he learned to decipher messages by ear, his bosses valued Andy as Pittsburgh's most accomplished telegraph operator, leading to yet another promotion. Tom Scott, the western supervisor for the Pennsylvania Railroad, lured Andy away from O'Reilly's as his personal telegrapher with the salary of thirty-five dollars per month—more than double his present earnings. The teenager started with the railroad on February 1, 1853, with the intention of learning everything possible about work and life.

Carnegie's father had taught him to dream. His boss, Tom Scott, became his mentor, teaching him to act. The Pennsylvania Railroad percolated with financial opportunity for those willing to nibble along the edges of morality. The Pennsylvania dealt with thousands of fees for cartage. The railroad had been organized along a quasi-military line with an emphasis on efficiency, ensuring all receipts remained in its coffers but offering inside information to Scott and Carnegie useful for personal gain. With sponge-like tenacity, Carnegie absorbed tidbits from Scott. The youth imbibed the mantra: "Mind the costs and the profits will take care of themselves."[21] He also aped his mentor's worship of the almighty dollar.

Carnegie vowed to be a doer. When a derailment torpedoed the system with Scott unavailable, Andy seized command, ordering the maintenance crews to clear the track via a memo signed under his boss's initials, "T.A.S." Traffic resumed to normal, and the trains lost little time. When Scott returned and discovered what Carnegie had done, he spoke with J. Edgar Thomson, the president of the railroad: "I'm damned if he didn't run every train in my name without the slightest authority," a breach in the command structure of the system. Since matters went well, Thomson commended Carnegie and called him "Scott's little white-haired Scottish devil."[22]

With the death of his father in 1855, Carnegie adopted Tom Scott as his surrogate father. Scott imbued him with the investor's mentality—money makes money. Scott provided specific financial advice as well, telling Carnegie to invest in Adams Express. Since the youth lacked the cash to make a buy, Scott loaned him the required $610. The dividends paid off the loan and earned Andy a tidy return.

When Pennsylvania Railroad promoted Scott as general superintendent in 1858, Carnegie accompanied his boss to Altoona as his personal assistant, with a nice increase in pay to go along with the new job. In Altoona, Scott's niece Rebecca Stewart polished Andy's manners, teaching him grace and etiquette as an addition to his business acumen. Carnegie donned high-heeled riding boots, sported a stylish black frock coat, improved the cadence of his Scottish lilt and perfected his storytelling. He purchased a handsome steed named Dash and honed his riding skill. He summoned his mother and brother from Allegheny City to live with him and hired a servant. Hard work, some luck and perseverance had brought him early success.

When Thomson advanced Scott to a vice-presidency, twenty-four-year-old Carnegie rose to superintendent of the western division in December 1859 at an annual salary of $1,500. Andy hired younger brother Tom as his personal secretary and telegrapher. The promotion required a return to Pittsburgh, the fog-shrouded hellhole whose clouds of smoke burned the eyes and stifled the breath. Carnegie wrote of the city in his *Autobiography*: "If you placed your hand on the balustrade of the stair it came back black; if you washed your face and hands they were as dirty as ever in an hour. The soot gathered in the hair and irritated the skin."[23]

D.A. Stewart, a Pennsylvania freight agent, suggested Andy purchase a home in the posh Homewood suburbs to escape the filth of the city. Carnegie's neighbors included the cream of Pittsburgh society—the wealthy Vandevort brothers, iron industrialist William Coleman and Judge William Wilkins.

Carnegie's populist principles garnered from his working-class family sometimes clashed with those of his snooty neighbors. When the judge's wife complained of "negroes" being accepted at West Point, Carnegie countered, "Mrs. Wilkins, there is something even worse than that. I understand that some of them have been admitted to heaven."[24]

The continuing Civil War had intensified the North's need for military transportation and communication systems, putting pressure on the Pennsylvania Railroad. President Abraham Lincoln summoned Tom Scott to Washington as assistant secretary of war to oversee telegraph lines and railroads. Scott drafted Carnegie to serve as his aide. During a repair mission in Northern Virginia, a loose wire slashed Carnegie's face, tearing open his skin—his red badge of courage. After shoring up the rail route between Alexandria and Washington, the "wounded" warrior rejoined Scott.

President Lincoln visited Scott's office daily, and Andy would later recall the commander in chief as one of the most homely yet charming men he had ever met. To provide the look of maturity to his appearance, the baby-faced Carnegie sprouted a blond beard to go with his long coat and high-heeled boots. After a four-month stint in Washington, he resigned his post to supervise troop and supply transportation from his Pittsburgh office as a Pennsylvania Railroad employee.

The war also had stimulated the country's need for kerosene and oil. Carnegie's neighbor William Coleman, whose daughters would wed Tom Carnegie and Tom Miller, invited Andy to tour the Storey Farm drilling site along the Oil River to witness the frenzy of the petroleum prospecting. The acrid smell of oil may have chafed Andy's nostrils, but his mind inhaled the rich aroma of greenbacks. Caught up in the speculative euphoria, Andy invested heavily in Columbia Oil, sharing the get-rich tip with his childhood "the Six" friend Tom Miller and boss Tom Scott. The first year's profits alone earned him $17,868.67. Additional investments in Woodruff Railroad Car, Duck Creek Oil, Dutton Oil, Pittsburgh Elevator, Third National Bank, Birmingham Passenger Railroad and Citizens Passenger Railroad shares made Carnegie wealthy. By 1863, his dividends generated more than $45,000 per year, equivalent to more than $800,000 today, contrasted against his Pennsylvania Railroad salary of $2,400.

The added work of wartime duties and investment management compounded Carnegie's job stress. His body and soul craved a rest. On June 28, 1862, Andy departed for an extended vacation to Scotland with his mother and Tom Miller, who stood half a head taller than he. Miller's fiery Irish temper contrasted with Carnegie's even disposition, but Andy admired his friend's keen mind. A serious bout with pneumonia downed Carnegie much of the trip. For nearly six weeks, he lay on a cot at the back of his Uncle Lauder's store in Dunfermline. Doctors bled and nearly killed him, but his young, strong body recuperated despite the medical incompetence. On the trip home, Miller received an emergency cable. Trouble was brewing at the axle plant. Anthony Kloman had begun drinking and carousing, which had set his more industrious younger brother Andrew on a tear. Upon Miller's return to the States, Andrew bullied him into buying Anthony's shares for $20,000, removing a thorn from his hide.

When Andrew Kloman realized the buyout left him with a minority position in his own company, he instantly regretted his hasty decision—especially after hearing the rumor that Miller had bragged his increased ownership was only the "camel's head within the tent."[25] His distrust magnified upon discovering Miller's earlier sale of shares to "the Six" debating club friend William Cowley without prior consultation. When Cowley died of typhoid fever during the Civil War, Miller had repurchased the stock for $8,500, again failing to notify his partner.

Andrew Kloman demanded that Miller sell him half of his brother's shares. To keep the peace, Miller sold him the half. Still not content, Kloman pushed Henry Phipps to sell his shares as well. "I can't do that," Phipps balked, carping like a nervous hummingbird to his protector Miller. "If I am put out, I shall spend the rest of my life keeping books for Dilworth."[26] Miller calmed Phipps: "Don't ye worry. Kloman's bark is far worse than his bite." Adding fuel to the fire, Kloman lashed out at Phipps for speculating on an oil contract that turned out poorly. Tension on all fronts intensified. When Miller withheld payment on a disputed Fort Wayne axle invoice from the company, Phipps turned on his defender for inhibiting cash flow—a cardinal sin to the money-minded treasurer. Compounding the brouhaha, a newspaper article mistakenly asserted that the corporate name would be changed to Kloman and Miller. On the advice of an attorney, Kloman inserted a renunciation in the August 20, 1863 *Pittsburgh Evening Chronicle*: "Notice is hereby given that Thomas N. Miller is not a member of our firm nor has any authority to transact business on our account."[27] The small lava flows encountered in most businesses had burst into a full-scale volcanic eruption. With discord spilling from the pot, the partners called in Andrew Carnegie as a mediator, who ruled: "Hearing both sides, I was fully satisfied I could not establish harmony upon the basis of a common partnership. I finally got all three together in my office and proposed that Miller should have his one-third interest and be a silent partner, Phipps and Kloman transacting the business. This was agreed to, but unfortunately, ill feeling was created."[28]

Kloman further demanded Carnegie incorporate a clause granting Phipps the option to buy out Miller for $10,000 upon sixty days' notice, a step Phipps promised never to take. Miller protested but inked the September 1, 1863 contract. The newly formed partnership increased the capital by $60,000:

Henry Phipps.

$30,000 from Kloman, $20,000 from Phipps and $10,000 from Miller. Since Phipps lacked the funds to foot his entire share, Tom Carnegie purchased half his shares.

Within months of the agreement, Kloman's paranoia festered. "I want Miller out and you must do it," Kloman ordered Phipps. Since Phipps feared Kloman more than Miller, he reluctantly agreed. However, Phipps lacked the required $10,000 to complete the option. Tom Carnegie again stepped into the fray, buying half of Miller's stock and loaning Phipps $5,000 for the other half. Miller carped to his friend Andy Carnegie: "He is a treacherous Judas, this Phipps. I was his friend, and he betrayed me. I might have expected a direct frontal assault from Kloman, but not this stab in the back from Phipps. Kloman may be an ignorant oaf without a likable bone in his body, but I can never forgive Phipps. That son o' a bitch will pay for his treachery." Such disloyalty surprised Andy Carnegie, even though he never could remain angry with his friend "Harry" for long. In a fit of pique, Miller formed a competitor, the Cyclops Iron Works, located just four blocks from Kloman and Phipps. Andrew Carnegie reluctantly joined Miller as an investor.

The new plant had been hastily planned and suffered a raft of technical inefficiencies. Tom Carnegie, a shrewd businessman, sized up the weaknesses and urged his brother to merge with the unstable but brilliant Kloman. Andy Carnegie recognized the wisdom of his brother's advice. Playing to Miller's ego, he urged him to overlook his grudge against Kloman and Phipps: "You return as a conqueror."[29] Miller softened and signed merger papers with the stipulation that he would avoid all contact with Phipps. Carnegie and Miller each anted up $50,000 along with their Cyclops assets to consolidate into the newly formed Union Iron Works. Andrew Carnegie became president; Tom Carnegie, vice-president; Henry Phipps, treasurer; and Andrew Kloman, production superintendent. Phipps steered clear of board meetings to avoid a confrontation with Miller. Peacemakers Andy and Tom Carnegie reassured

Phipps that the three of them held the votes to outvote Miller and Kloman on any significant issue.

With the Civil War further propelling iron prices to $130 per ton, the Union Ironworks rang up profits like a slot machine hitting three bars. Andy Carnegie and Tom Miller's railroad contacts proved invaluable, expanding the customer base. To support the growth in sales, the company purchased the Superior Mill and Blast Furnace. With outside investments dwarfing his Pennsylvania salary, Carnegie penned a March 28, 1865 resignation letter to President J. Edgar Thomson, who urged him to reconsider: "How does the position of assistant–general superintendent sound?" Carnegie declined. He intended to pursue his fortune as an independent entrepreneur, but first he planned an adventure. Carnegie departed Pittsburgh on May 17 for a grand tour of Europe with bachelors Henry "Harry" Phipps and John "Vandy" Vandevort. Tom oversaw the business in his absence and performed admirably, allowing Andy and his pals to jaunt through Europe. Good food and high living raised Carnegie's weight. "I am 7 stone, 11 pounds [109 pounds]," he wrote his mother.[30] The museums, galleries, cathedrals, popular landmarks and restaurants opened new horizons for the men. Henry Phipps marveled at the wonders of Europe, and Andy jotted the salient details in a journal.

The bachelor trio also explored English foundries and forges, finding the production technology far more advanced than that found in Pittsburgh. The new Dodd-Webb process for steel plating iron rails appeared especially promising. Upon their return to the States, Carnegie and Phipps tested the Dodd-Webb system on rail production, but the results failed the Pennsylvania Railroad's stringent standards. The experiment cost thousands of dollars and suppressed Carnegie's desire for future pioneering. "Stick to the tried and true, and you'll nare go wrong," he philosophized.

Andrew Carnegie possessed a magical knack for making money with minimal effort. Just one day prior to the European trip, he had established the Keystone Bridge Company with the backing of his cronies from the Pennsylvania Railroad, Thomas Scott and J. Edgar Thomson. Keystone quickly became the Union Mill's largest customer for iron railroad bridge beams. When President Edgar Thomson mandated that the Pennsylvania line replace all wooden bridges with iron for safety and durability, business roared full throttle.

The Carnegie-Scott-Thomson triumvirate farmed dollars in all directions. On April 9, 1867, they co-founded Keystone Telegraph, whose sole asset consisted of a contract to string telegraph lines alongside the Pennsylvania

Railroad. Although he had paid nothing for the company, Andy somehow employed his shifty smile, disarming Scottish brogue and questionable ethics to dance his way into $150,000 of Western Union shares in exchange for Keystone Telegraph. Each new Carnegie venture seemed to blossom like a tulip bud after a spring rain.

Miller still stewed over past treachery. He avoided Phipps and snubbed Kloman. When profits slowed at the end of the Civil War, Miller dumped his shares to Andy Carnegie for $73,600, giving Carnegie 39 percent ownership in Union. Despite his large stake in the company, thirty-five-year-old Andrew Carnegie held little interest in running day-to-day operations. "Tom is more than able to keep the ship afloat. I need a change of scenery." He handed the keys for his Pittsburgh home to his brother and relocated with Ma to a suite of rooms at the St. Nicholas Hotel in lower Manhattan, also establishing an investment office on Madison Avenue in partnership with Edgar Thomson. Carnegie would never again live in Pittsburgh. He would escape the hot New York summers at Cresson, a Pennsylvania mountain forest resort with B.F. Jones as his neighbor. The two iron barons rode, hiked, socialized and discussed business.

When Keystone Bridge landed the lucrative contract from the Keokuk and Hamilton Bridge Company to connect Illinois to Iowa across the Mississippi, Union received the iron order. Since Carnegie supervised the construction as well, he earned an additional oversight fee. Unfortunately, Andy forgot to advise his brother of the scheme. The iron order from Iowa Contracting crossed Tom Carnegie's desk without explanation. Tom glanced at the size of the order and advised secretary Gardner McCandless, "We should verify the company's credit worthiness." McCandless paused for effect and answered, "We consider the company good. We would add that such men as Messrs. J. Edgar Thomson and Scott are interested in the building of the road, and the treasurer of the company is a reliable New York gentleman, a Mr. Andrew Carnegie, with whose name you are perhaps not unfamiliar."[31]

Each new Carnegie venture had turned to gold. In yet another sweetheart deal, he converted Woodruff sleeper car shares into Pullman stock. However, Carnegie believed that profitable ends justified questionable means. His elfin appearance and Scottish charm made him irresistible to unsuspecting investors. When Keystone Bridge landed a $5 million contract for the Eads Bridge in St. Louis, the first use of large-scale cantilevered steel construction in the United States, Carnegie sold $1 million of the shaky bonds backing the bridge to Junius Morgan in London, receiving $50,000 in St. Louis and Illinois stock as a commission for his troubles.

When the bridge fell behind schedule, Carnegie needed interim financing. J.P. Morgan, Junius's blustery son, offered the funds, but on draconian terms: "I must warn you, should the bridge fail to reach completion by December 18, 1873, we will be forced to foreclose." Morgan exhaled from his omnipresent cigar and glanced toward the ceiling with the hint of a smirk. He thought Carnegie was such an arrogant buffoon. Carnegie thanked Morgan, promising, "The bridge will finish on schedule." He dodged disaster with but days to spare. Fifteen men died and seventy-seven were seriously injured during the construction of the world's longest arch bridge. As a stunt to demonstrate the bridge's strength, an elephant and trainer led a parade of excited admirers across the river. Despite the fanfare, the undercapitalized bridge ended in bankruptcy within the year, and Morgan would gain receivership, but not at Carnegie's expense.

Carnegie's wallet fattened as he stood on ever-slimmer moral ground. In one deal, he convinced Thomson to pledge Pennsylvania stock to underwrite a $600,000 bank loan for the teetering Union Pacific Railroad. In return, the railroad named Scott president and Thomson and Carnegie as board members. The Pennsylvania received $3 million of Union Pacific stock as collateral, and the trio received individual stock as well. With the fresh influx of cash, the stock immediately skyrocketed. Carnegie steamed to London in March 1871 with a briefcase full of the watered-down Union Pacific bonds to sell in the European marketplace. After pocketing the commission, Carnegie, Thomson and Scott dumped their high-flying stock at a tidy profit. When the stock plummeted, the bond and stockholders were left with the loss. Carnegie later apologized: "Instead of being a trusted colleague of the Union Pacific board of directors, I was regarded as having used them for speculative purposes. We were ignominiously but deservedly expelled from the Union Pacific board."[32] The expulsion was one of the few cases where someone caught Carnegie with his fingers in the cookie jar.

Carnegie had dealt in bad paper. He concealed information and misrepresented facts. While his construction companies profited, he double dipped, receiving hefty commissions on his bond sales. Swindled investors took Carnegie to court, a minor irritant. He usually won by dragging out litigation. On occasion, he repurchased defaulted bonds at near-bankruptcy pricing, re-hawking them at a higher price to other pigeons.

Massive wealth allowed Carnegie to take stock of his personal ethics. He found himself wanting. He had followed gold for gold's sake. With personal assets exceeding $400,000, worth perhaps $6,500,000 today, he vowed to up his moral compass. While sitting in his St. Nicholas apartment, he jotted with a stubby pencil: "The amassing of wealth is one of the worst species of idolatry. No idol is more debasing than the worship of money." He vowed to elevate his principles and become a better person, pegging a retirement in a few short years, just not that day. Money still sat on the table, and he intended to take it first.

• • • • •

While the iron barons feasted on the sweetness of rich cream, the poor laborers wallowed in the stink of stale milk—daily facing danger, disease, poverty and death for pitiful wages. Seeking a larger slice of the pie, the skilled boilers and puddlers had organized the Sons of Vulcan Union, establishing a "forge" at the American Iron Works on February 13, 1865. To combat higher negotiated labor costs, B.F. Jones devised a "sliding wage scale," floating wage rates up or down based on the selling price of iron. Since wartime prices were high, the union profited.

While these puddlers and boilers earned a decent living wage, the unskilled worker struggled to stay one step ahead of the bill collector. The contract system especially demoralized lifters and pushers. Management paid a fixed sum to a foreman who doled out pay to the men who worked beneath him. One bundler named William Labbitt skimmed $862.33 from a contract for $1,642.44, cheating both labor and management alike. Jones and the union worked together to eradicate the use of the contract system at Jones and Laughlin.

• • • • •

Ukrainian peasants suffered through an especially harsh 1863—the fall frost, a bruising summer, torrential spring rains and the winter freeze. Each day presented a host of routines and challenges. A peasant's farm existence proved difficult—up at dawn, feed the chickens, a quick breakfast of brown bread, off to school during the winter months. The chores, every day came the chores—chop firewood, clear the field, plant or sell beets for the few kopecks the family earned. The evening rarely ended later than nine o'clock.

While the Ukraine inched toward the Industrial Age, Russian invaders cracked down on individual freedom. In the larger cities like Kiev and Odessa, school already was taught in Russian, but not yet in the villages. The country witnessed the birth of its first railroad, but the modern world had yet to reach tiny villages like Nikolaevka. The motor was unknown, and plowing, planting and picking relied on horse or human power. Luckily, the Ukraine grew big strong boys. The farm offered no place for the weak.

Life was hard, but there were rewards. On Sunday after Mass, a roving baluska band might sing and play music in the square. After harvest, the peasants slaughtered a pig, and the entire village would feast on *salo*, cured slabs of fatback, to celebrate the crop. God had spared the townsfolk for another season, and most had enough to eat. Although the Russian oppressors controlled the cities, only a gentle rumble distressed the hamlets, farms and villages—at least so it seemed.

Over a dinner of hot borscht and boiled potatoes, one boy recalled the tears streaming down his mother's cheeks. Russian interior minister Peter Valuyev just had decreed that Ukrainian was merely a corruption of the Russian and Polish tongue and never existed as a separate language.

The decree meant the use of Ukrainian words no longer was permitted in schoolbooks—only Russian. The peasants also feared compulsory military service in the Russian army, which amounted to a virtual death sentence. Several boys had been conscripted over the past few years. None had returned. One mama quoted from her favorite Ukrainian writer, Ivan Kotlyarevsky, to her son: "The love to fatherland that is glorified no enemy force will withstand."

We can only imagine the mother's trembling fingers as she recognized the future hopelessness and touched her son's hand. She hugged him for several seconds just like she had done when he was a small child, the hint of a tear etching her eyes. Papa grasped his boy's hand and wished him well, knowing this might be a final goodbye. The father offered a few parting words: "You are a man now. Make your mama proud." A confused younger brother cried, "Do not leave me. Please stay." Too late; the decision had been made. The older brother would depart the farm for a new land and a new life.

With the few dozen rubles his mother pinned to the inside pocket of his peacoat, a bagful of clothing, a few loaves of freshly baked rye bread, a wrapper or two of cabbage-stuffed *varenyky*, a slab of dried pork and a hunk of cheese, the youth slipped out of his village at dawn and headed for the seaport of Odessa. With pounding heart, he felt alone. Should he be captured, his family would suffer. Most teenagers never had journeyed more than fifty miles from their villages. This youth had begun the arduous trek to a strange land—probably never again to return to the Ukraine or his family.

In the port city, the frightened youth purchased steerage passage on a freight sloop. He sailed across the Black Sea, the Mediterranean and the Atlantic Ocean on the six-week cruise to New York. Waves topping fifteen feet in height, rancid food and cramped quarters made the seas a nightmare. The Ukrainian's stomach churned, and his head tolled like a bell. He lay down, but the spinning worsened. He ate lightly, but his body refused to retain the food. He missed his younger brother, mama and papa. Everything seemed surreal.

The Eastern European peasant knew no English but practiced with a friendly Ukrainian sailor on calm days at sea. "*Pryvit*" in the native tongue translated to "hello." The English alphabet appeared strange—so unlike the Cyrillic. Mama would be proud. He would learn English. His cousin had called America the land of milk and honey in letters to his aunt. Someday, when his pockets were flush with dollars, he would look back on the journey as a great adventure when he described it to his children and grandchildren.

The New York coast of America offered a welcome, if intimidating, sight. The shoreline presented a hustle-bustle of activity. Burly blacks unloaded cargo bins onto wheelbarrows and wagons. Most Eastern Europeans had never before laid eyes on a black person. As the foreigner walked gingerly down the gangplank and off the ship, he felt like a tiny drone scurrying through a huge beehive. Words he could not understand in the immigration line confused him. The doctor checked his hair for lice and his body for disease, but the authorities eventually stamped his papers after several hours.

"Where go train?" he asked an agent in English like the sailor had told him. "Over there, Greenhorn," the man answered curtly, pointing to the railroad station. The crowded railcar mesmerized the immigrant. He never had seen a train, much less ridden on one. Anxious passengers hurried him into the car. The roar of the locomotive intermingled with the pungent smell of tobacco smoke, the sweat of the riders and the chattering of a foreign tongue. Once in a while, his ear recognized a stray word or two. The sailor had taught him "shit," which meant bad. He heard that word often. The vast woodland scenery passing by contrasted with the mass of humanity stuffing the train—fat men in frock coats, starched white collars and black cravats scanning newspapers; noisy boys in knickers scampering wildly along the aisle; a well-fed mother clutching a crying baby against her breast; even a sad-looking Croat or Serb, but no other Ukrainians with whom to share his discomfort. The conductor shouted out the name of each stop: "All out for Altoona." Nerves rattled his stomach. The sharp bark of the locomotive whistle and the constant swishing motion of the car made nodding off impossible. There was so much to take in—the flow of the rivers, the lush countryside and the buildings and farms bordering the tracks. How could he close his eyes? He might miss something.

A cousin greeted him at the rail depot and led him like a lost puppy to cramped quarters in a neighborhood of charred, nondescript row houses and rented rooms. A rickety flight of stairs snaked up to a third-floor landing. A splintery door begging for a coat of paint opened into a Spartan living room containing two chairs and a pine kitchen table. A well-used oil lamp served as the only decoration. A closet-sized bedroom contained a misshapen straw mattress, a bedside table with a few family photographs, a kerosene lantern and two candles. The quarters lacked plumbing. There was no soul. Where was the milk and honey?

Fresh from the old country, this hungry foreign-born farmboy needed a job—any job. Over a dinner of bread and leftover dried meat, the cousin described a part-time job at the Jones and Laughlin foundry. The pay would be bad, but at least the immigrant might learn a trade. The next morning, he would seek work, and the cousin agreed to translate.

The cousin joked in Ukrainian, "If you get job, you still poor. That mean you may only eat potato on Friday, the day before pay day. If you no get job, you dirt poor. That mean no potato on Friday." Both men laughed.

The cousin or friend would coach the new immigrant in his own tongue: "I talk. Pretend you understand. Whenever I look at you, just nod."

The two foreigners approached a swarthy Irish foreman with piercing eyes.

"What do you Hunkies want?"

"We ain't Hunkies. We Ukrainians," the more seasoned of the two would answer in heavily accented English.

"Hunkies, Cranians, what's the difference?"

"My cousin good man and strong. I work here. He need job."

"He don't say much."

"No, he quiet, but work hard."

"If I give him a try, it's one dollar per day. If he's late, we dock him. Late again and out the door. I don't want no back talk neither. Vershtay?" The recent immigrant would bob his head in silent agreement. He needed the part-time job in the foundry.

"Okay, let's sign you up. Over there by the table," he pointed.

B.F. Jones often kept track of his foremen. Sometimes, an immigrant might spot the splendidly attired man with a long flowing beard in a suit and tie by the table and mistake him for the recruiter. "You mean that man?"

"No! That's Mr. Ben Franklin Jones. He owns the place. You want the guy in the flannel shirt and black cap with the sideburns."

The city of Pittsburgh, circa 1907.

Immigrant laborers worked hard. They unloaded hoppers and shoveled coke into the furnace, swept floors, toted stock and hefted ingots—hour after hour, day after day. Their bodies acclimated to the heat, the noise and the stink. The bosses fired the weak and complainers. When their arms ached and their muscles burned, these men escaped to another place, another time—mama, papa, the farm. The smartest adopted a mantra: practice English—learn more, earn more. They now were Americans. English proved difficult. A select few studied the dog-eared dictionaries their landladies might own but possibly couldn't read. Others picked up the slang of their co-workers, today's Pittsburghese. The most diligent learned to read, finding newspapers and an occasional book that took them from the slums or the mill floor to the world of imagination. Those who had not yet lost hope still believed anything was possible in America.

• • • • •

While many foreign-born laborers struggled through part-time menial jobs at Jones and Laughlin and other small foundries, Andrew Carnegie honed his iron-making management skill at the Union plant by installing the planning and organizational programs he had learned from Tom Scott at the Pennsylvania Railroad. He explained:

> *As I became acquainted with the manufacture of iron, I was greatly surprised to find that the cost of each of the various processes was unknown. I insisted upon such a system of weighing and accounting being introduced throughout our works as would enable us to know what our cost was for each process and especially what each man was doing, who saved material, who wasted it, and who produced the best results.*[33]

America had entered the Industrial Age. Rapid advancements in science and technology would soon change the dynamics of the iron industry forever. Pittsburgh was about to leap full tilt into the Age of Steel.

CHAPTER 2

J. EDGAR THOMSON

William Coleman took a sip from a cup of sweet tea and glanced at his son-in-law across the table. He raised his eyes toward the ceiling and pontificated: "Those of us in the iron industry who fail to act on the new opportunity in steel will rue the day." Tom Carnegie nodded his head in agreement. He did not need to be a metallurgist to recognize the obvious advantages in the durability of steel over iron.

The Pennsylvania Railroad already had initiated a program to replace iron with steel rails because the alloy promised to hold up for years instead of months. During the past decade, a handful of specialty steel manufacturers had tiptoed into the marketplace. Anderson and Cook had built the Pittsburgh Steel Works during the Civil War era. The Black Diamond and Crescent Steel Works followed shortly thereafter. These pioneers hired Welsh and English Sheffield-plate workers to manufacture high-quality tools using the crucible technique, by which flux, a cleaning agent, was added to melted blister bars to purge impurities and produce small batches of high-quality but expensive steel.

Modern steel manufacturing received a jump-start in 1851 following Pittsburgh native William Kelly's experiments demonstrating oxygen's chemical affinity to carbon. A blast of hot air shot through a cylindrical chamber of molten iron in Kelly's foundry reduced the percentage of carbon and generated "refined" iron, what we now call steel, about 60 percent of the time. Kelly sold most of this new output for steamboat boilers. In 1870, the

Cambria Iron Works in Johnstown employed Kelly's patented "pneumatic-air-boiling process," earning its developer $30,000. William Kelly would die a wealthy, respected steel expert in 1888.

Others simultaneously experimented with the refinement of iron into steel. On August 11, 1856, Englishman Henry Bessemer delivered the landmark paper "The Manufacture of Iron Without Fuel." The Bessemer Process pushed hot air into a sealed, pear-shaped vessel of boiling pig iron, engendering a violent chemical reaction. A rainbow of multicolored sparks pyramided more than thirty feet in the air. An onrush of dark-brown fumes and dull-red flames called a "blow" roared like a banshee, signaling the birth of a steel solid as it separated from the manganese and silicon slag that remained behind as scrap.

In 1857, inventor Robert Forester Mushet discovered that spiegeleisen or "looking-glass iron," an iron, carbon and manganese alloy, reduced brittleness and blowholes in steel, problems that had baffled both Kelly and Bessemer. Spiegeleisen further enhanced quality by substantially removing rogue sulfur, developing the final ingredient in a perfect recipe for low-cost, mass-produced steel.

Captain Eber Ward purchased the Bessemer steel process for his Detroit plant as early as 1865. Brown University–educated engineer Alexander Holley designed another facility in Troy, New York, and an even more advanced acid works near Harrisburg a few years later, both heavily financed by J. Edgar Thomson's Pennsylvania Railroad. William Coleman had visited several of the new Bessemer plants and admired the elegance of the new technological advances.

With his father-in-law's prodding, Tom Carnegie convinced his brother to investigate the Bessemer process. Andrew Carnegie relaxed in an easy chair and listened to Tom. He weighed the pluses and minuses. Steel-plated rails had burned him a few years earlier, but his brother laid out a strong case for steel. Pioneering could be painful, but he risked the loss of the Pennsylvania Railroad's orders if he did nothing.

The cooling of wartime demand along with the introduction of steel already had slowed iron sales. While Tom Carnegie struggled to soothe the massive egos of the Union Iron executive team, chief financial officer Henry Phipps rowed through Pittsburgh's economic waves to balance a sea of bills with reduced receipts. When short on cash, he paid workers with script redeemable for food at a local grocery store. Almost weekly, Phipps drove a black buggy drawn by his mare Gypsy from bank to bank in search of temporary funding. Gypsy followed the route by memory: a stop at Citizen's National, the First

The Bessemer Process pushed hot air into a sealed, pear-shaped vessel of boiling pig iron, causing a violent chemical reaction.

National and eventually to Third National Bank on Wood Street. Phipps faced the steely bankers hat in hand with a sad-sack face, obtaining loans to float a check for two or three days until the required cash arrived.

Phipps assiduously executed the management techniques he learned from Carnegie to boost output and control expenses. Detractors accused him of dealing unscrupulously with his scrap vendors, claiming he used twin wagons to cheat. Supposedly, one wagon weighed five hundred pounds more than the other. Accusers stated he weighed the first load both empty and full using the heavier wagon. On subsequent trips, he substituted the heavier wagon with a lighter one that would be weighed full, shorting the unsuspecting vendor by five hundred pounds. One poor Scottish peddler complained, "Divil a cint was left to a hard wurrrking man after a thrade with Henry Phipps." His never-ceasing vigilance over dimes and nickels and his willingness to handle Carnegie's dirty work made him an indispensable asset. The treasurer's nervous energy supplied the stamina to hunt for squandered assets like a ravenous mouse prowling for a chunk of cheese. Mere table scraps could cater a sumptuous banquet for Phipps.

In 1872, Carnegie traveled to Europe toting a briefcase crammed with $6 million in watered-down Davenport and St. Paul Railroad bonds. Although the bonds eventually defaulted, he foisted the securities on unsuspecting investors and earned a $150,000 commission in the process. During the trip, Carnegie took a side stop to Suffolk, England, where he toured the Bessemer furnaces. Witnessing the low-cost manufacture of big-batch steel instantly converted him to a Bessemer Process adherent. Scottish cousin and Glasgow University–trained mechanical engineer George "Dod" Lauder confirmed his opinion, and Carnegie inked a licensing agreement with Henry Bessemer on the spot.

Carnegie's Pennsylvania Railroad connections virtually guaranteed the success of a large-scale steel rail plant in Pittsburgh. After huddling with Phipps, Coleman, Kloman, brother Tom and a handful of well-heeled investors, Carnegie anted $250,000 toward the projected $700,000 initial capitalization for a Bessemer mill.

"I found the perfect spot for our mill, but it will be expensive," Coleman advised his partners. He had located a site in Braddock, a Pittsburgh suburb where General Edward Braddock had lost his life during the French and Indian War. The 107-acre property bordered the Monongahela River and the Pennsylvania and Baltimore and Ohio Railroad tracks. The proximity to coal, limestone, iron ore, rail and river transportation and an industrious labor force pushed the odds even higher.

The new firm of Carnegie and McCandless incorporated on November 5, 1872. The name of bank officer David McCandless, a fellow member in the Swedenborg Church with Carnegie, provided the proper touch of respectability. McCandless, Tom Carnegie, Henry Phipps, Columbia Oil president David Stewart, Andrew Kloman and Allegheny Railroad officers John Scott and William Shinn each subscribed for $50,000 of stock. William Coleman, Tom Carnegie's father-in-law, signed on for $100,000. Carnegie appointed Shinn as general manager at an annual salary of $5,000, permitting him to retain his railroad position as well.

Plans moved ahead swiftly. Treasurer Henry Phipps negotiated the sale of the original Union Lower Mill to Wilson, Walker and Company, with Andy Carnegie retaining a minority position for his own personal account. In the past, Carnegie had diversified his investments. Now, he intended to specialize in steel. He wrote in his *Autobiography*: "Put all good eggs in one basket and watch the basket."

Carnegie christened the plant J. Edgar Thomson in honor of his previous employer, the president of the Pennsylvania Railroad and the top prospect for future business. In return, Thomson received a small stake in the mill. As a quid pro quo, the Pennsylvania granted the mill inside shipping rebates. Unfortunately, Thomson would die in 1874, just prior to the grand opening of the plant.

Some of Carnegie's contemporaries questioned his chances for success. Industrialist John Moorhead whispered to an acquaintance as Carnegie passed him by on the street: "There goes a foolish young man. He has bitten off more than he can chew. He wasn't satisfied to do a small safe business like the rest of us. Mark my words, he'll come to grief yet."[34] Carnegie would prove Moorhead wrong.

Carnegie knew how to attract top talent. He hired engineer Alexander Holley—the holder of fifteen patents, author of more than three hundred magazine articles and designer of eleven of the country's first thirteen Bessemer plants—to create the plans for the facility. Carnegie explained, "I am neither mechanic or engineer, nor am I scientific. I seem to have had a knack of utilizing those who knew better than myself."[35] Convinced he had delegated construction to capable hands, he, his mother, Mr. and Mrs. William Coleman and Mr. and Mrs. David McCandless departed in February 1873 for a two-month riverboat holiday, keeping in touch with Braddock by letter and telegraph.

Upon his return from vacation, he was forced to face the reality of the Panic of 1873, which decimated the economy over the next several months. The Jay Cooke and Company Bank went belly up, and the New York Stock Exchange closed for ten days to stem the bloodbath. Pittsburgh suffered with the rest of the country. Judge Thomas Mellon, president of T. Mellon and Sons, summed up the recession: "The vitals of trade were destroyed by the canker worm of credit until the decayed carcass dropped dead."[36] Several of Carnegie's partners reneged on their cash commitments. The Exchange Bank threatened to terminate its line of credit. Although the economy had thrown a sharp curve, Carnegie sold Pullman and Western Union Stock to obtain needed funds, stretching his finances to the hilt.

The panic delivered a financial shock to Tom Scott as well. Greed propelled him into an untenable construction loan. Andy had begged his mentor to steer clear of the investment, but Scott persisted. Now, he begged Andy to co-sign for nearly worthless Texas Pacific Railroad notes as collateral to stave off bankruptcy. Carnegie knew his signature would drag him down along with his friend. Thomson implored Carnegie to help, writing, "You of all others should lend your helping hand."[37] In his *Autobiography*, Carnegie wrote that denying his friend "gave me more pain than all the financial trials to which I had been subjected."[38] Watching Scott slowly sink in a swamp of debt-filled quicksand cost him a friendship but saved Carnegie from certain ruin.

The economy's decline provided positives as well as snags. Tempestuous Andrew Kloman had speculated in a shaky iron ore investment that had backfired, forcing his sale of shares to Carnegie at bargain-basement pricing. The financial squeeze also caused Coleman to sell back his shares at par.

To obtain the funds necessary for the completion of the plant, Carnegie sold $400,000 in construction bonds to British financier Junius Morgan. Years earlier, the Morgan family had owed Carnegie $10,000 on a railroad investment and $50,000 as a bond commission. When Carnegie claimed his $60,000, Junius Morgan's son Pierpont handed him a check for $70,000, explaining the overage as an understatement in accounts. "Will you please accept these ten thousand with my best wishes?"[39] Although the two men later developed a distaste for each other, Carnegie appreciated Morgan's honesty and vowed never to harm him.

The $1,250,000 cost for the J. Edgar Thomson plant exceeded budget, but a weak economy allowed Carnegie to employ personnel at lower-than-planned wages. He picked top-quality talent, painting a bright picture for his young geniuses—growing foremen, superintendents and managers from within the organization. "Mr. Morgan buys his partners. I raise my own," he boasted.[40]

• • • • •

Lines of desperate men, many newly arrived immigrants, snaked around the gate of the new J. Edgar Thomson plant prior to opening. Poles, Czechs, Slovaks, Serbs, Croats and Hungarians queued up, anxious for work. Dozens applied for every job. The foremen selected only the biggest and strongest at a pay rate of seven dollars per week with the reminder, "We don't take kindly to no troublemakers."

• • • • •

Captain William Jones, an assistant to Alexander Holley, represented Carnegie's ultimate hiring coup. Born in 1830 to a poor Welsh pattern maker in Catasauqua, Pennsylvania, not far from Allentown, Jones distinguished himself by earning a captain's battlefield commission during the Civil War. At the end of hostilities, this squat, muscular leader with a slight stutter joined Cambria Steel in Johnstown for two dollars per day, rising to the number-two job in the company. He quit in a huff when passed over for the superintendent's slot.

At Edgar Thomson, Jones invented a host of laborsaving devices saving hundreds of thousands in costs—most notably the Jones Mixer, a monster-sized, brick-lined iron blender with a 500,000-pound capacity. Carloads of trains dumped melt into the mixer, where the gentle rocking motion blended the mix into uniform output.

Steelmaking proved a veritable house of horrors. Boiling steel dangled precariously in giant cupolas above the heads of the workers, hissing like a coiled snake ready to strike. Escape valves seethed with lethal steam, threatening the careless. Train cars overloaded with coke and lacking protective guardrails dropped slabs on unprotected limbs. Long hours and heavy lifting made for exhaustion. Tired men got hurt. Accidents occurred—slips, falls, drops, torn muscles and death.

Day in, day out, those at the plant faced a malevolent jumble of machinery and furnaces, cognizant of the perils involved. They performed their duties cautiously. Unfortunately, injuries were a numbers game, and each week, fate somehow drew some unlucky soul's number. If a defective chain from an emptied cupola snapped from its coupling and flew across the floor, some grunt might die—savaged by the malfunctioning equipment. Broken arms, cracked ribs, crushed fingers and sliced limbs occurred with frequency.

The cry "Gurney" set the plant astir. Those uninjured thanked God for their safety.

"Who's down?" a man might question.

"Is he dead?" another might ask.

"Looks bad, real bad."

Broken arms were set at the hospital. Stitches came without anesthesia. Infection might kill since doctors had little understanding of its severity. Survivors might hear, "You are lucky to be alive. Had that chain struck a few inches higher, you would be dead." Survivors returned to the mill quickly. They needed the paycheck.

• • • • •

J. Edgar Thomson produced its first steel rail on September 1, 1875. Big-batch steel employed large and expensive machinery. The mill's three Spiegel cupolas, measuring five feet in diameter by forty feet in height, melted the pig iron, which was remelted under pressure in the massive five-ton Bessemer converter. Workers emptied the resultant steel into casting ladles, where it was molded into fourteen-inch-wide ingots, each weighing nearly one ton. Power-feeding tables sliced the ingots into seven-inch blooms. A three-ton hammer further downsized the blooms into three sections that would each become a rail. A twenty-three-inch, three-tiered rail mill rolled out a thirty-foot rail every two minutes. Six burly laborers hooked the sixty- to sixty-five-pound rails once they passed through the rollers. Profit-driven supervisors and foremen ensured the plant hummed at near capacity—two hundred tons of rails per day. Although rail prices had declined, low manufacturing costs more than compensated for the drop. While competition charged seventy dollars per ton, Carnegie billed steel rails at sixty-five dollars per ton, generating handsome returns due to the plant's low-cost technology.

Master marketer Andrew Carnegie excelled at prospecting for sales. He lobbied the 1876 Philadelphia Centennial Exposition to construct half its main building from J. Edgar Thomson steel rather than wood as a "fire-safety" precaution. The Carnegie-controlled Keystone Bridge Company naturally received the lucrative contract to construct the building's skeleton.

To gear up pig iron production for the increasing demand, the company constructed the Lucy Furnace on Fifty-first Street, named in honor of Lucy Coleman Carnegie, Tom's wife. A second furnace built in 1877 upped output even more.

Captain Jones invented an automatic driving table to feed rails mechanically, supplanting the labor-intensive "hook-and-tong" process. The table reduced roller employment from fifteen to five. Each improvement by Jones cut the need for skilled labor. Although a handful of higher-paid men lost their jobs, displaced by technology, the foreman promoted a select few to Edgar Thomson's expanding furnace department. A contract for the Brooklyn Bridge's steel girders necessitated the replacement of the five-ton Bessemer converter with a larger seven-ton unit.[41]

With Captain Jones directing manufacturing, Phipps tending to the books and Tom Carnegie, William Shinn and David McCandless running day-to-day affairs, Andrew Carnegie felt confident in his team. Frequent reports kept him up to date. With matters flowing smoothly, he departed for a yearlong European holiday with his mother and "Vandy" Vandevort, publishing a record of his adventures, *Notes on a Trip Around the World*. Partway through the holiday, he received a telegram with the unsettling news of the death of Chairman David McCandless, which initiated a scramble for the COO's spot.[42]

When the plant's general manager, William Shinn, discovered that Andy planned to promote his brother Tom instead of him, he interviewed for the top position at Vulcan Iron in St. Louis. The plot thickened when Carnegie discovered Shinn owned part of Peerless Company, a J. Edgar Thomson limestone supplier. Carnegie, no stranger to insider deals, viewed the arrangement as a direct conflict of interest. Shinn resigned under a cloud of suspicion in September 1879. Carnegie ordered the redemption of Shinn's stock at book value, far below its true worth. Shinn accused Carnegie of "willful and malicious mendacity" and threatened legal action.[43] The parties called in iron veteran B.F. Jones as an impartial arbitrator but failed to reach a compromise. Only when Shinn instituted a suit in the U.S. Circuit Court did Carnegie settle at a higher price.

Plant superintendent Captain Jones had made J. Edgar Thomson into the most successful steel plant in the industry. After a day at the mill, he purged his work stress by playing the piano at his home or relaxing on the porch with a good five-cent cigar. As he marched into McDevitts Emporium on his way from work for a few stogies, a peppy young clerk greeted him with a welcoming smile.

"Need three cigars, young man." Jones placed fifteen cents on the counter.

"Yes, sir," the teenager answered smartly, laying the cigars next to the change. He looked up at his customer. Everyone in Braddock knew Captain William Jones. "Hot day," the clerk offered with a smile, picking up the coins and dropping them into the till.

"A very hot day, but it's hotter in the foundries," chimed in the Captain.

Charlie Schwab steeled up his courage and shot a remark to the flinty superintendent: "Yes, but I'd rather be in the works than here—a lot. And if you could only give me a job I would be so thankful."[44]

The Captain asked the clerk to slow down a little but listened as the youth described his aspiration to become an engineer at J. Edgar Thomson. Jones admired the lad's spunk, reminding him of his own unbounded ambition some three decades earlier.

Charles Michael Schwab was born on February 18, 1862, in tiny Williamsburg, Pennsylvania. His parents had struggled to earn a living but spared Charlie little. They scrimped and saved to buy him a piano and music lessons. The Schwab family life centered on Charlie, a born entertainer who loved to sing, act and perform. The boy's memory for names and faces and ability to organize others impressed all who met him. As a child, he convinced his younger brother Joe to shovel snow with him in the neighborhood for a few pennies.

At age twelve, Schwab moved to Loretto, where his father hauled mail. In school, he stood at the top of his class through a combination of effort, reasoning and moxie—but most of all a gift for gab. His teacher Father Ambrose Laughlin pegged him well: "Charlie was a boy who never said: 'I don't know.' He went on the principle of pretend you know and if you don't find out mighty quick."[45] He completed his education at St. Francis, a two-year Catholic college, which provided a standard high school curriculum combined with introductory college-level perspective drawing, surveying, engineering and bookkeeping.

Charlie lived near Cresson, where metal titans like B.F. Jones, Andrew Carnegie and other rich Pittsburgh and Johnstown residents spent their summer vacations. Of Carnegie he wrote, "I little thought at that time, when I held his horse and did trivial services for him, that fate in later years of life would so intimately throw our lives together."

Folks liked "Smiling Charlie," including the ladies. Shortly before his seventeenth birthday, he met Mary Russell, an aspiring Pittsburgh actress on a visit to her sister in Loretto. The two fell in love and planned to elope. When Mary's family discovered the affair, they shipped her back to Pittsburgh, crushing the lovesick teenagers. Charlie moped around the house for weeks until his parents came up with a plan to make him forget Mary. Charlie's mother purchased a store-bought suit, pinned a five-dollar bill inside a pocket and shipped him the sixty-five miles by train to Braddock for a clerk's job at McDevitt's Emporium. From 7:00 a.m. until 9:00 p.m., Charlie stocked shelves, waited on customers and swept the floors.

"This humdrum job bores me," he confided to co-worker and friend Tom Watson. The afternoon Captain Jones arrived at the store provided an opportunity he ill could afford to waste. Schwab may have exaggerated his credentials to earn a job at J. Edgar Thomson, but he learned quickly. Charlie later reminisced, "If I had not sold that five-cent cigar to Bill Jones, I might still be selling dried apples over-the-counter."[46]

Schwab started at J. Edgar Thomson on September 12, 1879, at one dollar per day as a rod man, driving stakes and toting leveling gear for the layout of furnace installations. While assisting the surveyors, he absorbed the basics of steel manufacturing from borrowed metallurgy and engineering books during evening studies. "I had a desire to know all I could."[47] His hard work paid dividends, and the Captain helped him move up the line.

When the plant temporarily assigned chief engineer Peter Brenslinger to its iron ore mines on the outskirts of State College, Schwab took over as interim department head. Upon Brenslinger's return, the Captain moved

Charlie to the drafting department to broaden his engineering experience. When a full-time promotion became available, the Captain devised a test. He scheduled overtime work on a project without additional pay. He ordered the chief draftsman to tell him which men performed the best and without complaint. Jones received just one name: Charlie Schwab.

Charlie performed admirably at each position he held. One day, the Captain summoned his apprentice: "Hey, Charlie, I'm busy. Take this report to Mr. Carnegie. If he asks any questions, tell him about things here. You know them as well as I do." "Yes, Captain," answered the wunderkind. While waiting in Carnegie's parlor, he spotted a piano. He sat down and played a ballad, humming in accompaniment. As the Scot entered the room, Charlie stopped and began to apologize for making himself at home. "Don't stop on my account. Play a Scottish ballad."[48] Carnegie loved music, and once he ascertained Schwab's steel manufacturing savvy, Charlie became a favorite. Jones liked music nearly as much as his boss. He hired Charlie to teach his daughter piano. Charlie sent most of the fee he received to his parents in Loretto as a thank-you for his own childhood lessons.

Within six months, Schwab had advanced to assistant plant manager. Jones continued as his mentor and "best friend of early life."[49] Charlie even collaborated on the development of the Jones Mixer, and each completed assignment gained him added recognition with upper management.

Schwab remained in touch with Tom Wagner, his former co-worker at McDevitt's. Tom had married and lived at his mother-in-law's boardinghouse. While visiting the newlyweds, Charlie sat down at their piano and played a tune. As he whistled the melody, Emma Eurania Dinkey, a chunky and not particularly attractive girl, cozied up beside him. Charlie smiled at her and introduced himself. "And they call me Rana. I am Tom's sister-in-law," she answered.

The improbable couple kept company for the next three and a half years—he a Catholic, she a Presbyterian; he outgoing and she quiet; he two years her junior in age. When Charlie proposed, Rana joked, "Why do you want to marry an old lady like me, lad?"[50] The couple wed on May 1, 1883, honeymooning in Atlantic City. Upon their return, they moved into a modest but comfortable wood-framed cottage near the plant. For the next fifty-five years, Rana would call Charlie "Lad" and he would call her "Old Lady."

Charlie Schwab had come to the steel industry during a remarkable period of growth. When prices increased to eighty dollars per ton, Superintendent Jones upped monthly raw steel output to twelve thousand tons at a cost of thirty-six dollars per ton. By 1880, annual rail production had jumped to seventy-six thousand tons. Carnegie pontificated, "A perfect mill is the road to wealth."[51] "It is only the beginning."[52]

Schwab recognized the opportunity before him and threw himself into his work, up by 5:00 a.m. for an early start. He championed scientific management despite the Captain's complaints: "Damn it Charlie, chemistry is going to spoil the steel business yet."[53] Charlie disagreed. He usurped Rana's sewing room and converted it to a mini laboratory. Each night, he analyzed steel samples for phosphorous and carbon content and checked corrosion resistance levels. When Carnegie partner Henry Phipps heard about the experiments, he volunteered $1,000 for additional equipment.

• • • • •

The injured laborers who healed from the ravages of accidents wore their scars like symbols of survival. Other workers generally ignored the disfiguration. They knew they might just be the next victim. A thick mustache might camouflage facial damage. Little could be done to disguise a limp, a deformed hand or open gaps from missing teeth.

The most ambitious pursued advancement, wanting more than a common laborer's position. A rumor of an impending retirement or a new job opening allowed those with the greatest seniority to apply. Discrimination generally disqualified Eastern Europeans from contention. Only those with a decent command of English received a shot. If the timing proved propitious, those with the nerve to ask and the necessary communication skills might push their case, sometimes receiving an apprenticeship to gain a foothold into a semi-skilled operation. If the immigrant obeyed orders, kept his mouth shut and expressed no union agitation, he could receive the position. With the plant growing and more furnace and boiler operators needed, a foreigner might advance if the bosses could overcome their bias against Ukrainian, Hungarian or Polish birth.

Finding a wife proved another challenge for Pittsburgh's immigrants. The number of workingmen seeking brides in Pittsburgh far outnumbered available single women. A family introduction, a meeting at church or even blind coincidence might blossom into marriage. If the worker lacked connections, he would feel like a ship lost at sea with no compass or a port in sight. One Ukrainian immigrant's story began with a walk to a neighborhood grocery store to buy vitals for his landlady—eggs, salt, bread, cheese, apples and some turnips or potatoes. With a smoky haze blackening the sky and winter's fury in full force, the wind seared his cheeks. The dirt streets oozed muddy slush. Like many of his compatriots, the worker was a good Catholic who believed God had some purpose for him. Like Job, this lonesome soul accepted freezing weather and bad events with faith in a better life ahead—just three blocks to the store to warm his hands and get the groceries for his landlady.

The shivering steelworker opened the creaky store door. The drafty room welcomed him. The man spotted a young lady standing before his eyes—a beautiful angel in a long plaid flannel skirt and dark red blouse. Her raven-black hair and hazel eyes brought shivers to the love-starved worker. He lacked family to help with an introduction and had reached his mid-thirties. This fine lady appeared nearly a decade his junior. She was beautiful and out of his league. He was disfigured by a mill accident, a large scar across his face.

"May I help you?" asked the girl in halting English, as she had been trained to do.

He ordered a chunk of cheese, a loaf of bread and a box of salt. He then remembered the potatoes. He thought there might be something else, but he could not remember. All he could think about was the girl. He paid her for the supplies and bounced out the door with thoughts of her beauty dancing through his head. As he reached the walkway, he spotted a young boy playing with a ball on the street, oblivious to a rapidly approaching horse and wagon. Surely the boy would hear the driver and step aside, but he took no heed. Just as the horse and wagon came within inches of crushing the child, the man reached out and snatched the boy to safety.

At that very moment, the shop girl opened the door in search of her brother. Where had he gone? He had promised to play by the steps. Just as she looked toward the street, she witnessed the customer save the child. She ran to the street and grabbed her brother, unsure whether to hug or punish him.

"Thank you, sir. I saw what you did for my brother. He is deaf and could not hear click-clack of wagon." She introduced herself. Her accent displayed the pleasant high-pitched tinkle of a Christmas bell.

And thus by fate began the unlikely romance between a Ukrainian steelworker and a Polish girl—an oasis of joy in a desert of despair.

• • • • •

Shrewd, strong-willed Margaret Carnegie had molded son Andrew in her image. Purpose surged through his veins, permeating every nerve. He personified the drive his father lacked. Andrew pointed to his mother's forehead and told an admiring acquaintance: "Here's where Tom and I got our brains."[54] In Andy's mind, no other woman could match her. He doted on "Ma" and vowed to remain single as long as she lived. Although an eligible bachelor, his mother tied him to her apron strings. Then, Andy met Louise Whitfield.

Fellow Scotsman and threads dealer Alexander King asked Andy to accompany him for a condolence call to the Widow Whitfield and her three children on New Year's Day 1880. Her daughter Louise was twenty-two, twenty years Andy's junior, but he was in the prime of life with only a hint of gray flecking his blond hair and beard. A ruddy complexion and pleasant grin made him appear younger than his years. Louise was a bright woman who spoke fluent French, had traveled extensively, possessed impeccable manners and stood several inches taller than Andy.

Andy and Louise talked of horses and their favorite places to ride. When King invited Louise to ride with him in Central Park later in the week, Carnegie accompanied them. Andy displayed strength and confidence on the saddle, and Louise likewise proved a skilled rider. A few days later, Andy asked Mrs. Whitfield's permission to take her daughter on an un-chaperoned ride. The two spent a delightful afternoon together, and the worldly industrialist won over the young girl's heart. Louise jotted in her diary: "After my first ride, I decided, whatever the future had in store, that would remain the great experience of my life."[55] The frequency of the outings increased, and the seriousness of the May-December romance progressed as the couple discovered their shared interest in reading and riding. Carnegie presented Louise with the book *The Light of Asia* by Edwin Arnold—the allegory of Prince Siddhartha's search for a rare pearl. The avid readers spent hours discussing the novel's symbolism.

With the love affair blossoming, jealousy seized Margaret Carnegie's heart. Now seventy, she determined no "chippie" would put her claws into Andra' if she could help it. Andrew had promised his mother a summer holiday in Scotland for the groundbreaking celebration of the library he donated to his hometown of Dunfermline. Andy intended to include Louise along with the Phipps and King families, which included other single women in a party of "gay charioteers." Margaret told Louise's mother, "If she were a daughter of mine, she wouldna go."[56] Louise did not go, but she never forgave "Mag" Carnegie for her interference, calling her "the most unpleasant woman she had ever known."[57]

A parade of the grateful populace of Dunfermline greeted the Carnegies, who gloried in the adulation. "Is this not all I promised and more?" Andy asked his mother. "Aye, it's grand," beamed the old lady. The crowd roared with admiration when Margaret laid the memorial stone for the magnanimous gift, which contrasted with her son's parsimony to his own labor force. At the June 19, 1881 dinner celebrating "Queen Dowager" Margaret's birthday and the founding of the library, Carnegie reminded the assembly that his own father and a handful of weavers had formed the city's first book-lending collection decades earlier. He was proud to extend the Carnegie family's love of books to the good people of Dunfermline. A local dignitary rose and lauded the Carnegies' benevolence. Mrs. Carnegie smiled broadly. When the spokesman jokingly mentioned that "only a lass for Andy" could make the evening more complete, Margaret's face darkened. Andra' was hers. Carnegie memorialized the wondrous journey in the travelogue *Our Coaching Trip*, which he dedicated to "My Favorite Heroine, My Mother."[58]

• • • • •

The typical immigrant steelworker could ill afford the luxury of travel. Since escaping the old country, few ever had left the confines of Allegheny County or journeyed more than twenty miles from their front doors. These unfortunates envied the high life of their bosses—the trips to the capital cities of Europe, the fine houses and the fancy carriages. They never would walk arm in arm with their ladies to the tony boulevards of Fifth Avenue in New York or cruise across the Atlantic to European playgrounds.

• • • • •

J. Edgar Thomson's steel business continued to thrive through 1881. Andrew Carnegie consolidated the Union Mills, the Lucy furnaces and J. Edgar Thomson into Carnegie Brothers and Company, Ltd. He owned $2,737,977.55 of the $5 million capitalized stock. His brother Tom and Henry Phipps each held $878,098. John Scott and John Vandevort had a portion of the balance. In a further move toward expansion, the company constructed two larger pig iron furnaces, divesting the Lucy units to Wilson, Walker and Company, a smaller firm partially owned by Phipps's brother-in-law.

Uninterrupted flows of iron ore, limestone and coke—the most important basic ingredients required in steelmaking—proved critical to maintain production efficiency. Shortages of materials created bottlenecks—an anathema to management. When a coke outage shut down three furnaces one afternoon, the superintendent sent the entire team home, cursing the idiots in purchasing. The stoppage cut into company profits and aggravated the labor force, which squawked at the loss of work time.

Carnegie Brothers and Company, Ltd., owned coke facilities but not enough to ensure a steady supply. At the next board meeting following the outage, Phipps pointed out that a competitor could purchase coke on the open market far cheaper than from the company's internally owned ovens. "Such a situation is intolerable. We must team up with a substantial coke partner," voiced another board member. "My sentiments as well," seconded Tom Carnegie. After a lively discussion, Tom targeted the H.C. Frick Coke Company, a vendor that dominated Connellsville's high-grade coke market, as a likely candidate. Its president, Henry Clay Frick, known as the "King of Coke," possessed a solid reputation as a serious, tough performer.

Henry Clay Frick.

Henry Clay Frick and his lovely new bride, Adelaide, were honeymooning in New York. Carnegie invited the young couple to brunch with him and his mother in his hotel suite. After the requisite chitchat and a pleasant meal, Carnegie blurted out an invitation for an alliance. Margaret, who joined her son and frequently offered her own unsolicited advice, hissed in his ear: "And what's in it for us?" Andy's answer must have satisfied her, as plans for a merger quickly followed.

Frick and Carnegie formalized a partnership during an 1882 meeting in Scotland. Carnegie had done his homework. Every ton of pig iron required 1,700 to 1,800 pounds of coke for fuel.

Carnegie's move had tied up the best coal lands in the region, hundreds of beehive ovens for the conversion of coal to coke and signed on a ruthless CEO with a "positive genius for management."[59]

Frick and Carnegie made an unlikely duo from the start. Both stood barely five feet, three inches. Carnegie sported a salesman's gift for gab. He relished the lilt of his own voice. He quoted rich poetry from Burns and Shakespeare to stress a point. On occasion, he tweaked the truth or skirted along the edge of morality to win his way. In contrast, Frick acted with directness. This strategist and organizer listened carefully. He only spoke when necessary and always meant what he said. Superintendent Captain William Jones summarized the two men: "I don't particularly like Frick, but you always know where you stand. With Carnegie it's a different matter. He is a sidestepper."[60]

Both men maneuvered for absolute control of the coke company, which Frick considered his personal fiefdom. Carnegie looked at the whole picture, interfering whenever high coke costs raised steel prices. Frick, a natural sourpuss, grew even pricklier each time Carnegie intruded. From the onset, the two clashed like vinegar and oil.

When Carnegie balked at a proposal to purchase additional Connellsville coal lands, Frick stood firm, winning his way. This early disagreement would by no means be the last. With steel production skyrocketing, the coke syndicate, which included H.C. Frick Coke, mandated an increase from $1.15 to $1.50 per ton. Frick supported the move. A cost-driven Carnegie countermanded the increased coke rate, undermining his lieutenant's authority. A stunned Frick bristled at the affront, addressing his coke cohorts without emotion: "Gentleman, you just heard what the worthy representative of the majority stockholder in the H.C. Frick Company said. Now if you will excuse me, I shall take my leave."[61] He coolly departed but vowed never again to allow his senior partner to embarrass him.

• • • • •

Charlie Schwab's career as assistant superintendent at J. Edgar Thomson moved forward at breakneck speed under Captain Jones's mentoring. In 1885, the twenty-three-year-old newlywed designed a railroad bridge to carry molten iron across the Monongahela River, successfully completing the project below budget and ahead of schedule. The Captain rewarded him with a diamond-encrusted spider pin and a compliment: "Great job as always, Charlie." Carnegie added twenty gold pieces. When Charlie escorted his father to the project, John Schwab beamed with pride, providing the most valuable reward of all.

Based on a solid recommendation from the Captain, Schwab received the assistant superintendent's spot at the newly purchased Homestead Plant under Superintendent Julian Kennedy in 1888. The workforce of J. Edgar Thomson had liked Schwab and hated to see him leave. He led by example, treated the men well and knew steel. When Charlie performed up to expectations at Homestead, management promoted him to superintendent following Kennedy's retirement, earning him the rich salary of $10,000 per year.

The J. Edgar Thomson plant ran as nonunion. Under Andrew Carnegie's direct order, the company chairman, William Abbott, initiated a boot-out-the-union campaign at Homestead during the final months of the contract, scheduled to end on July 1, 1889. Superintendent Schwab did his part and quelled a full-scale mutiny by the workers. Carnegie wrote, "Glad Schwab proved so able."[62] Unfortunately, Abbott caved and traded a three-year contract with reduced wage rates for labor peace. The lack of backbone cost Abbott his job. Tom Carnegie had died in 1886. Henry Phipps wished to retire. Captain Jones had no interest in the job, and Schwab lacked experience. Carnegie replaced Abbott with Henry Clay Frick.

The untimely death of Captain Jones in September 1889 forced Carnegie to make several critical executive moves. Twenty-seven-year-old Charlie Schwab replaced the Captain as superintendent at Edgar Thomson. John Potter took over Schwab's position as superintendent at Homestead. On October 10, 1889, Chairman Frick formally introduced Schwab to the staff. Schwab nervously whispered to his clerk, William Powell, "I can hardly realize that I am general superintendent of this plant. What do I know more than you fellows about this business?"

The men on the floor welcomed Schwab back to Thomson. They knew he would make the plant hum, just like the Captain had done, and high production meant steady work. When a foreman proved incapable of motivating his nighttime crew. Schwab asked, "How many heats has your shift completed?" "Six," the man answered. Charlie wrote a "6" on the floor and challenged the men to beat it. The next day, the "6" was erased and replaced with a large "7." The next shift crossed out the "7" and replaced it with an "8." Charlie had increased productivity with minimal effort.

Schwab maintained an open-door policy for his staff, and he spoke with the workers when he passed the furnaces to feel the temperament of the men. He could be tough, very tough, but fair. During grievance proceedings, he invited the committee members to sit, passed out cigars and chitchatted about the weather or current events to dissipate tension before getting down to business.

Charles Schwab.

Schwab faced serious labor trouble at the plant on New Year's Eve 1890. Earlier in the day, the J. Edgar Thomson stockyard team had walked off the floor over a furnace worker wage dispute. When sympathizers joined the protest, production halted. After drinking up their courage at Wolfe's Bar, an inebriated mob of sixty so-called Hungarians vandalized the plant, cracking the heads of the security team who stood in their path. Schwab rallied an opposing posse armed with clubs and formed a protective wall surrounding the furnaces. The confrontation lasted for several hours until the defenders drove the rabble from the plant.

The following day, the labor army reappeared in increased numbers. The confrontation began with worker threats and name calling and ended in minor bloodshed before the invaders retreated. Luckily, neither side encountered serious injury. Fearing another foray, Schwab armed one hundred men with Winchester rifles and prepared for the worst, penning a note to Chairman Frick: "I understand the Huns intend to make another attack tonight, and I can assure you that if they do they will meet with a pretty lively reception. I am determined to drive them out, no matter at what cost."[63] Schwab's forces carried the day. The resurrection fizzled, and the defeated men begrudgingly returned to work.

Charlie Schwab immersed himself in steelmaking, relishing the perks that came with his plant manager's position. He took to the good life like a fish in water. He hired famed architect Frederick Osterling to design a handsome red-brick mansion decorated with slate-roofed towers, which admirers likened in elegance and size to Chairman Henry Frick's stately mansion, Clayton. A separate stable housed his horses. To thank his parents for their support, Schwab purchased them a $300 yearly annuity. He also donated to the construction fund for Braddock's St. Thomas Catholic Church.

Charlie adhered to the scrap-heap philosophy of his former boss, Captain Jones—reduce variable costs through technology. Throw out the old; bring in the new. A pressure pump costing $5,000 significantly amplified power and paid for itself in short order. The purchase of a $21,638 three-tier, forty-inch blooming mill outpaced rival Illinois Steel's manufacturing capabilities. By 1891, J. Edgar Thomson controlled 19.13 percent of the country's entire rail production. Frick voiced a resounding approval for Schwab's work performance: "The men are extremely pleased with the new order of things, and it is considered by everyone that the works certainly are in better shape than ever." Andrew Carnegie called him his "Number 1 Superintendent," an affirmation of his ability to squeeze costs, increase output and maintain labor peace.[64]

During his early years, Henry Phipps had etched his own mark with an accountant's red pencil by lopping and chopping, slicing and dicing. He considered himself the voice of fiscal reason in the Carnegie Empire. He tried to rein in the company purse strings by opposing Schwab's more extravagant equipment expenditures. His naturally conservative nature had advanced with age, as he preferred a dose of higher dividends to the reinvestment of profits. He wanted to enjoy his money. This limited-growth strategy rubbed Schwab the wrong way. The superintendent believed a company that rested on its laurels risked eventual annihilation, and he considered Phipps a "little dandified man who was not assertive."[65]

Lately, Carnegie had inoculated Phipps with the philanthropy bug. In early 1890, Mary Schenley donated a park to the city. As Andy Carnegie and Henry Phipps coached through Schenley Park accompanied by Pittsburgh public works director Edward Bigelow, the three discussed a proposed $50,000 gift from local businessman Charles Clarke for a conservatory similar to the one Phipps previously had given to Allegheny City. Carnegie chimed in with his own two cents: "Harry," as he always called Henry, "you ought to build a conservatory right here. You're not going to let Charlie Clarke get ahead of you, are you? You can afford four times $50,000 and give Pittsburgh a fine conservatory." Carnegie knew the project would enhance the nearby library and museum complex he had donated. Phipps rose to the bait. "I'll do it," Phipps blurted, and do it he did. After decades as a penny pincher, Phipps finally displayed a giving soul, and his philanthropy grew with the years.

Phipps gifted $25,000 to Pittsburgh's library to open on Sundays. When the clergy crucified his stance as ungodly, Phipps countered, "It is all very well for you, gentlemen who work one day in the week and are masters of your time the other six during which you can view the beauties of nature, but I think it is shameful that you should endeavor to shut out from the toiling masses all that is calculated to entertain and instruct them during the only day which they have at their disposal."[66]

• • • • •

The workers knew the plant owners only through reputation. They were far removed from the common day-to-day mill hands. A laborer's contacts generally were limited to the plant foremen or, occasionally, plant superintendent Schwab. Ben Franklin Jones and Henry Oliver, both dressed to the nines, had toured J. Edgar Thomson to inspect the furnaces. Andrew Carnegie, Henry Phipps, George Lauder and Henry Clay Frick paraded around the floor in suits and ties from time to time to show their faces, but their presence did nothing to improve the everyday pain endured by the workers. A pot-bellied fireman shot sparks toward the executives as they passed by him. A tall Serb mouthed profanities, or at least it seemed so. One crude Hungarian, lacking the gift of English, spit on the floor as soon as the owners walked out of sight.

• • • • •

Andrew Carnegie displayed enormous pride in his executive staff, which was composed of mostly younger men. When U.S. president William Henry Harrison toured the Carnegie plants following the dedication of the Carnegie Library in February 1890, the Scot introduced his crew, one by one. After the president shook hands with Charlie Schwab, he turned to his host and asked, "How is this, Mr. Carnegie, you present only boys to me?"

"Yes, Mr. President, but do you notice what kind of boys they are?"

"Yes, hustlers, every one of them," Harrison answered. That was Carnegie's secret—men like Schwab in boys' bodies.

Despite his respect for Schwab, Frick frequently clashed with his superintendent on personnel and financial issues. The two were polar opposites. Cold, calculating accountant Frick buried his feelings, while outgoing Charlie spoke his mind. As top man, Frick required control. When Schwab pushed for his brother-in-law's promotion, Frick abruptly squashed the appointment with "He's not ready yet." On another occasion, Schwab procrastinated on a mandated dismissal. Frick came down hard: "It seems very difficult to get anything into your head or have you follow instructions as given you. I wanted Mr. Bullion's services dispensed with. I shall expect prompt action on your part in this matter."[67]

Schwab flew by the seat of his pants, providing broad-brush explanations. Frick required specific budgetary proof. "The next time you are in I should like you to call on Mr. Lovejoy [the company secretary] and let him show you the amount of money already spent at Edgar Thompson this year for improvements. I think it will make you open your eyes," cautioned Frick.[68]

Despite their occasional disagreements, both labor and management respected Smilin' Charlie Schwab, a man who expected results but had a nice way of getting the men to do what he wanted. He could be forgiving. One winter afternoon, a laboratory worker pelted him with a snowball. Charlie fired the offender on the spot. The man apologized. Schwab relented, informing the foreman with a grin: "Take him back, but explain to the darn fool that I can't have every man hitting me with snowballs."

J. Edgar Thomson flowered under Schwab's charismatic leadership. He combined "Frick's greed, Jones's charm, and Carnegie's taste for empire with a better grasp of statistics, chemistry, and metallurgy," wrote one historian.[69] Charlie enjoyed his stay in Braddock at J. Edgar Thomson, but a strike at Homestead soon would change everything.

CHAPTER 3

HOMESTEAD

A contingent of financially strapped investors approached Andrew Carnegie and Henry Clay Frick hat in hand during the fall of 1883. They desperately needed to dump the Homestead steel plant, located a mile away from J. Edgar Thomson. Former Carnegie partner and steel expert Andrew Kloman had designed the plant, but he died prior to its completion. Without Kloman's firm leadership, labor unrest and weak management had drained the new mill's productivity. With desperation consuming the Homestead stakeholders, the Carnegie team smelled an easy win. Frick offered a price of $350,000 with a small allowance for the increased land value. The hungry investors seized the offer. All but one took cash. Only William Singer accepted Carnegie Brothers stock, a move that would one day make him exceedingly wealthy. Labor relations at the new purchase challenged the Carnegie team, but the plant paid quick dividends.

The triumvirate of Tom Carnegie, William Abbott and Henry Phipps handled day-to-day business decisions while Andrew Carnegie spent most of his time in New York, Cresson or Scotland reading, writing magazine articles, fishing, golfing and supporting pacifism. He invested in a chain of radical English newspapers that pursued better working conditions for the British workingman while he continued to ignore the plight of his own American laborers.

Andrew Carnegie prided himself on spending little time in his Pittsburgh plants. His knack for getting to the nub of a problem allowed him to peruse sales and production data from home and provide a list of suggested actions

Homestead steel works.

to the executive team. When a local businessman bragged of being in the office from seven o'clock in the morning until dark, Carnegie mockingly countered, "You must be a lazy man if it takes you ten hours to do a day's work. What I do is get good men. I get reports from them. Within an hour I have disposed of everything, sent out my suggestions. The day's work is done, and I am ready to go and enjoy myself."[70]

In Europe, Carnegie hobnobbed with the famous, numbering Parliament member and secretary to Ireland John Morley, poet Matthew Arnold, his brother novelist Edwin Arnold, biologist/philosopher Herbert Spencer and British prime minister William Gladstone among his close associates.

Most found Andrew Carnegie to be a curious sort. Some thought him brash and loud. The snooty, political Reform Club of England blackballed him. Yet he charmed Lord Rosebery of Scotland, and New York's prestigious Union League elected him to membership. He displayed the skill of a master negotiator but frequently employed his economic clout to score a win. He was tight with his pennies at the plants but gave millions for philanthropy. Vain, quick to criticize, a bully and a flatterer, he may have been a flawed man but was nonetheless a great one. One critic wrote, "There is not one Andrew Carnegie; there are really a half dozen."[71]

While Andy Carnegie gallivanted across the globe, brother Tom remained the glue holding the company in place. He mitered the joints at all the right places, soothing the massive egos of the executive team. William Abbott backed up Tom Carnegie. Captain William Jones, prior to his untimely death a few years later, made manufacturing hum while brooking no guff. Phipps guarded the pennies. Frick ran a profitable coke company, brooking Andrew Carnegie's interference, which irritated like sand in an oyster. Foreign-born laborers pushed out the product with minimal protest.

The Carnegie management team detested unionism, following the tenet: "Might makes right." In December 1884, Captain Jones posted a sign at Edgar Thomson: "Plant Closed for Equipment Installation." In actuality, the plant shut down and starved the Knights of Labor and the Amalgamated Union's 1,600 members into submission.

• • • • •

Getting by proved difficult for the typical millworker, always just one misstep from the poorhouse. A family might hoard dimes, dress frugally and eat modestly but barely eke out a living. One laborer confessed to his priest: "They crush us in a vise and not give us break, making it hard to put food on table. These are hard men without Jesus in their hearts, but what can we do? We must work as told."

• • • • •

The month of January 1885 proved especially cold. The shortage of cash had squeezed the J. Edgar Thomson workers into submission, forcing them to sign individual contracts for a twelve-hour day at reduced hourly wages in exchange for meat and potatoes on the table and money for rent after a month's inactivity without pay. The plant would remain nonunion for nearly two decades. With labor under foot, toiling at subsistence wages, Carnegie spent the following summer at Cresson penning *Triumphant Democracy*, his book extolling America's greatness and the opportunities for individual growth in a free land.

In Greek tragedy, hubris or pride often leads to the downfall of the great. Calamity struck Andrew Carnegie with both barrels in 1886. On July 21, he collapsed from flu-like symptoms. As he healed, serious illness assaulted both his brother Tom and mother, Margaret. Still weakened from his undiagnosed malady, typhoid fever struck him in October. Bedridden at his Braemar cottage in Cresson, he teetered on the brink of death. Forty-three-year-old Tom Carnegie, debilitated by disease, alcohol abuse and his brother's constant haranguing, died of pneumonia. His mother passed away within weeks on November 10. Andrew lay comatose, too ill to be told of his mother's passing in the next room. Andrew Carnegie's physician, Dr. Frederic Dennis, secreted Margaret's coffin from the cottage so as not to disturb his patient's precarious recovery.

Typhoid fever struck the working class as well, killing hundreds during that hard year. Relatives and friends of the deceased prayed to God for the strength to steel themselves against the tragedies of life.

By early December, Carnegie's health had slowly improved. With his body still mending, he departed Cresson for New York, saddened by the loss of his brother and mother. Under the care of physician Dr. Jasper Garmany, he spent two months convalescing at Dungeness, his sister-in-law Lucy's two-thousand-acre estate in Cumberland, Georgia.

With his mother and brother laid to rest and his health returned, the fifty-one-year-old bachelor realized how short life could be. He wedded Louise Whitfield on April 22, 1887, at her mother's home on Forty-Eighth Street, seven years after their first meeting. Less than thirty attended the evening rite, which included close friend and business partner Henry Phipps along with his wife, Anne. Prior to the ceremony, the bride, who patiently had awaited her wedding day for many years, signed a prenuptial affirming: "The said Andrew Carnegie desires and intends to devote the bulk of his estate to charity and educational purposes and said Louise Whitfield sympathizes and agrees."[72] In return, Andy gifted Louise with a portfolio of stocks and bonds yielding annual dividends of $20,000, as well as a $200,000 mansion on Fifty-First Street, neighboring the Vanderbilt estate, which he purchased for his wife prior to the wedding.

The couple departed by coach for the steamship *Fulda* for an extended honeymoon through England and Scotland an hour after the ceremony. Louise proved a wonderful wife who idolized her husband's philanthropy and overlooked his foibles.

• • • • •

Steel workers sought escape from the mill through small luxuries. "I not work tomorrow," one Ukrainian worker from J. Edgar Thomson bragged to his family. "It is Fourth of July. We have day off. Men at plant talk of new courthouse opening. Hundreds go. We go and take baby." This would be a special treat. The worker families boarded the downtown trolley, joining the crowds of well-wishers for the courthouse opening. The massive stone Gothic exterior looked like the stone buildings he once had seen in Odessa. The "Bridge of Sighs," the partition that separated the courtroom from the jail, reminded the immigrant of his own escape from Russian military servitude. A tear dotted his eye as he remembered his dear deceased mother and the money she had pinned inside his coat. He thanked God for his wife and son as he grieved for the old country and family he would never see again.

• • • • •

Homestead had continually experienced a history of Amalgamated Association of Iron and Steel Workers (AAISW) union slowdowns, work stoppages and intrusions into management's prerogatives, an anathema to production-oriented Carnegie. He instructed his executive staff to oust the union with the completion of the labor contract ending July 1, 1889—just as he had done at Edgar Thomson a few years earlier. He offered every employee an individual wage rate. Labor naturally defied the order and struck. On the tenth, the Allegheny County sheriff escorted a carload of scabs to the mill. Militant strikers blocked their entry. Carnegie ordered President William Abbott to employ a rock-hard, no-quarters-given stance.

Faced with union pressure, Abbott buckled and traded a three-year sweetheart contract at a reduced pay scale for labor peace. The move so infuriated Carnegie that he fired his CEO on September 7 with the criticism: "It can never be accepted by your partners that you can't compete with and whip others. This is what our young partners are partners for."[73] Abbott's dismissal opened the race for a successor.

Regardless of his public statements on the right to organize into unions, Andrew Carnegie, like Jones, believed the average worker lacked the capacity to handle money beyond what was required for the basic necessities of food, clothing and housing. He intended to invest in grand schemes like libraries, concert halls, church organs and museums that would stand as perpetual monuments to his name and benefit all mankind rather than improve the welfare of the individuals who struggled to make ends meet.

On September 28, shortly after releasing Abbott, Carnegie completed his latest charitable project, writing in his *Autobiography*: "Yesterday,

I strolled out with Henry Phipps and walked over to see the library in Allegheny. If ever there was a sight that makes my eyes glisten, it was this gem. I saw many people standing and gazing and praising the big words, Carnegie Free Library."

Pittsburgh's city council originally rejected Carnegie's offer of $250,000 to build an even larger library in Oakland, objecting to the requirement for $15,000 in annual upkeep. Once the city fathers eyeballed the handsome new Allegheny City library, they reconsidered. When representatives asked if the offer still stood, Carnegie declined with a impish grin—counter-proposing a larger $1 million gift but upping the annual-upkeep demand to $40,000. City council accepted, and he felt good about himself.

Tom's death and Abbott's firing created a huge hole in the business hierarchy. Board members Henry Phipps and David Stewart quietly captained the ship through rocky shoals on a temporary basis, but Stewart's death and Phipps's desire for retirement from daily operations eliminated them as candidates. With limited options, Carnegie tapped Henry Clay Frick to head up Carnegie Brothers in addition to his presidency of H.C. Frick Coke. The strong-willed Frick intended to stamp his image on Carnegie Brothers just as he had done on the coke company.

• • • • •

Workers groaned under brutal twelve-hour-days at J. Edgar Thomson. The overworked crew struggled with torn muscles, dehydration, aches and pains. Even hard-ass Captain Jones had supported a ten-hour day. Weak men grew careless. Long hours heightened danger. Again, the men walked off in protest, but this time Carnegie executives answered with a shutdown while they retooled equipment and furnaces.

A plant delegation railed to New York for direct negotiation with their Scottish boss. Carnegie welcomed the negotiators like lost children, regaling them with tales of his early days in steelmaking. He handed out an article he had written and introduced them to a visiting Siamese dignitary but stonewalled any compromise on hours or working conditions.

Starved into submission, the cowed workers returned to J. Edgar Thomson under Carnegie's draconian terms. The men performed like zombies, beaten stares lining their haggard faces, a sour taste in their mouths—no smiles, no jokes. Injury statistics climbed. "This not God's will. What kind of man treat us like slaves? My brothers and I deserve better fate," complained one weary immigrant soul.

• • • • •

Frick commanded the Carnegie Brothers ship with efficiency while Carnegie luxuriated in Scotland, France, England and Italy. He gloried in the perks of running the country's largest steel manufacturer but cringed with every public pronouncement his outspoken boss made to the media, unsure whom Andy might offend or embarrass next. At least, the time Carnegie spent writing kept him out of Frick's hair. The June 1889 issue of the *North American Review* featured Carnegie's article "The Gospel of Wealth," lauding planned charitable giving by the rich as a benefit for the general populace. He moralized to his millionaire contemporaries: "The man who dies rich dies disgraced."[74] His wealth and communication skills gained him the political clout to earn a seat at the Pan American Conference and successfully power the nomination of Pittsburgh attorney George Shiras into a Supreme Court seat. Carnegie joined fellow industrialist B.F. Jones in lobbying successfully for the McKinley Tariff, which provided a $13.40 per-ton duty on rails and $20.16 for structural shapes, virtually eliminating foreign competition.

Although Carnegie's political and public connections delivered business, his nitpicking irritated those who dealt with him. Captain Bill Jones called his boss "a wasp that came buzzing around to stir up everybody." One muggy afternoon, Carnegie approached his superintendent and apologized: "Captain, I'm awfully sorry to leave you in the midst of hot metals here, but I must go to Europe. I can't stand the sultry summer in this country. You have no idea that when I get on the ship and out of sight of the land what a relief it is to me." The Captain exhaled a blast of smoke from his trademark cigar and smirked, "No, Andy, and you have no idea what a relief it is to me either."[75]

When the Duquesne Steelworks entered the marketplace as a competitor, Carnegie publicly criticized its steel for a lack of "homogeneity." Following Frick's successful purchase of the company on November 21, 1890, the homogeneity problem mysteriously disappeared, and J. Edgar Thompson switched to the direct-rolling process, the very technology Carnegie had criticized.

Sheer size of operation and equipment superiority provided the Carnegie plants with a winning hand. "The market is mine when I want it," Carnegie bragged.[76] The wily Scot played his political contacts and melded his media publicity as trump cards. Membership in the right clubs and organizations provided access to the rich and powerful. He explained, "Big contracts are always more likely to be made over nuts and wine than across a desk."[77]

In January 1891, Louise Carnegie contracted typhoid fever. Andrew Carnegie insisted Dr. Denis care for her daily. Not until April 3 would she regain the strength to venture outside her New York home. During his wife's recovery, Carnegie engrossed himself in the construction of his newest philanthropic project, Carnegie Hall. On opening night of May 5, the audience stood and applauded as a healthy Louise and her husband took their seats to hear composer Peter Tchaikovsky. Carnegie greeted the public acknowledgement of his gift with a smile and a slight bow. He loved the attention of the crowd.

Even while Louise Carnegie battled typhoid fever, Frick faced his own turmoil. At nine o'clock in the morning on January 27, 1891, a massive blast rocked the H.C. Frick Company's Mammoth Shaft in Scottdale, sending 110 Eastern European miners to eternity. A stray spark from a mine lamp ignited gas fumes, burning dozens of bodies beyond recognition. Family men Martin Shavinsky, Andrew Vashol and Andrew Charensky joined unmarried Jacob Shongon, Steve Yatsk and Michael Kosinsky in unmarked mass graves at Mount Pleasant's St. John's Cemetery. One family member bemoaned, "They didn't even give them boys no tombstone."

• • • • •

Despite the horrific accident, the year 1892 started strongly for the Carnegie companies. The plants operated near capacity, churning out product and profits. Homestead initiated construction of a mammoth naval armor plate mill, largely at the government's request. The military theorized that whoever controlled the seas ruled the world. A strong naval force required steam-powered, steel-plated, armed warships. Secretary of the Navy Benjamin Tracy pushed to develop a United States armada. Competitor Bethlehem Steel had won the contracts for naval steel plate, but it had fallen behind. Tracy pressured Andrew Carnegie to have Homestead take up the slack.

To tighten up the company structure, the board authorized the merger of Carnegie Brothers and Carnegie Phipps into Carnegie Steel, Ltd., effective July 1, 1892. The consolidation joined J. Edgar Thomson, Homestead, Duquesne, Hartman Steel in Beaver Falls, the Upper and Lower Mills, the Lucy furnaces and the Larimer and Youghiogheny Coke Works into a single entity. Only Carnegie-controlled H.C. Frick Coke would remain separate. Henry Clay Frick served as COO of the corporation. Charlie Schwab ran Edgar Thomson, and John Potter oversaw Homestead. Senior board members Dod Lauder and Henry Phipps supplied maturity and reined in runaway expansion. Carnegie, owner of 55 percent, gleefully exclaimed, "Was there ever such a business?"[78]

Three years earlier, Abbott had failed in the attempt to toss out the AAISW, whose membership included only 800 of Homestead's 3,800 employees. Since then, advanced technology had altered steel economics. The new open-hearth ovens and the advent of electricity had reduced the need for skilled labor. Illinois Steel had cut its prices on rails from thirty dollars to a March low of twenty-three dollars per ton. Carnegie Steel required a corresponding price reduction to remain competitive. As the Homestead contract neared its end, management presented a take-it-or-leave-it 15-percent wage reduction to its skilled workers, a bitter pill the AAISW could ill afford to swallow. As an additional barb, lead spokesman Frick demanded all future contracts terminate in the heart of winter on December 31, a slow time for production, rather than June 30.

In a year that had started so well, the men at nonunion J. Edgar Thomson questioned how far Frick intended to squeeze its sister plant. The defeat of the AAISW at Homestead would carry terrible repercussions for them as well. The entire steel industry watched the Homestead contract negotiations with rapt anticipation as each side angled like children maneuvering for leverage on a seesaw. The workers at Edgar Thomson conversely prayed for a pro-union miracle. A loss would create a terrible setback for all labor. A victory might even rekindle unionism at J. Edgar Thomson.

Frick knew the union would reject the 15 percent reduction. His true motive was to eradicate the AAISW forever, not just reduce wage rates. As early as the April 4 board meeting, Carnegie floated the idea of posting a notice reading: "As the vast majority of our employees are non-union, the firm has decided that the minority must give way to the majority. These works will be necessarily non-union after the expiration of the present agreement."[79] The board argued that so aggressive an offensive might lead to counterproductive results. Carnegie agreed, granting Frick full autonomy over negotiations.

Regardless of the company's grousing about declining rail prices, the workforce at Homestead believed their greedy bosses were growing fat on steel plate and structural beam business. Labor, especially the unskilled, wanted higher pay, not a reduction. On the other hand, management only could see its massive investment in the latest equipment. They questioned why labor should share in the profits. Each side dug in, setting the scene for the most violent labor-management confrontation in Pittsburgh history.

On April 13, Louise and Andy Carnegie sailed for an extended holiday at Ranoch Lodge in Scotland. Henry Phipps, as board spokesman, urged Carnegie to remain in Europe and refrain from public comments. "The welfare of the company required that Mr. Carnegie should not be in the country, because he was disposed to grant the demands of labor, however unreasonable."[80] Carnegie bit his tongue and vowed not to interfere. He would allow Frick to run the show.

Frick surrounded Homestead with a six-foot-tall protective picket fence dubbed "Fort Frick" by labor. Carpenters sharpened each board to a point and drilled holes in strategic locations just large enough for rifle barrels. Frick instructed Superintendent Potter to deliver an ultimatum: workers must accept twenty-two dollars per ton for billets instead of the current twenty-five dollars. If they refused, Homestead would operate as nonunion. Backed into a corner, the union negotiators rejected further negotiations. Labor forever would consider Henry Clay Frick as the reincarnation of Satan.

• • • • •

Few knew Henry Clay Frick, this man whose name forever would be tied with Homestead. He buried his inner feelings beneath a veil of silence. Industry executives respected his domineering personality and organizational panache. Tycoons such as John D. Rockefeller and J.P. Morgan liked his style. The grunts in the plant hated him. With the accountant's knack for getting to the heart of a problem, Frick sliced through clutter to extract the right solution. Although he looked soft and friendly, he could be hard—very hard. Despite a small stature, he called the shots. Like Carnegie and Phipps, Frick never reached five feet, four inches. Art dealer René Gimpel summed up the duality of his personality: "His features are so regular, his face so pleasant, that he seems benevolent, but at certain moments you could see and comprehend that you were mistaken. His head is there, placed on that body, for his triumph and your defeat."[81]

Clay, as most called him, was born in West Overton on a gray morning, December 19, 1849, a Wednesday's child, full of woe, named for the renowned Kentucky senator Henry Clay. By nineteenth-century standards, folks considered his maternal grandfather, Mennonite Abraham Overholt, land rich. The family patriarch owned 260 acres of grain-producing property, as well as top-selling Old Overholt Rye distilleries in Overton and Broad Ford, southeast of Pittsburgh. Overholt spurned Frick's father, John Frick, as a ne'er-do-well, a rough-and-tumble miller and a Lutheran to boot. Only the premarital pregnancy with Clay's mother forced Overholt's begrudging acceptance of so unsatisfactory a son-in-law.

Clay lived in the springhouse with his parents on his grandfather's farm and endured a sickly childhood. Scarlet fever felled him during a visit to his paternal grandparents, who lived in Van Buren, Ohio. Weakened by a near bout with death, he lacked the stamina to compete in sports, but he excelled at chess, checkers and horseshoes and played the fiddle. Although shy and socially awkward, he carried himself with dignity and performed well in school. His ashen hue, soft features, jutting chin and diminutive size masked mental toughness. Silent, unsmiling and methodical, he displayed the orderliness of an accountant in an unprepossessing physical shell. Yet he remained fearless, standing up for his rights, even if it meant taking a beating at the hands of a larger classmate.

Grandfather Overholt set a high bar for his grandson, who intended to rise above it. Clay expounded, "Everyone was put in this world to work There is nothing to hinder me from being worth one-million dollars."[82] After completing a course of study at the Western Pennsylvania Classical and Scientific Institute, Clay attended Otterbein College in Westerville, Ohio, for six weeks. He rounded out his early business resume by keeping the books for the family at the Broad Ford Distillery and clerking part time at an uncle's store.

Clay's Uncle Christian connected the teenager with a full-time job at Eaton's general store for six dollars per week. In short order, he landed a better spot at linen and lace dealer Marcum and Carlisle for eight dollars per week, rapidly advancing to the store's top producer and a salary of twelve dollars per week. During breaks, he slipped upstairs to Schuchmann's Lithography, directly above the store, where he nourished a budding love of art. In the evening, he took an advanced accounting course at the Iron City Commercial College.

Eager to leave behind the drabness of his Mennonite upbringing, Clay borrowed fifty dollars from his Uncle Christian to trade in his rural clothes for a suitable business wardrobe. He joined the Fourth Avenue Baptist Church and summoned the nerve to ask his grandfather, "Won't you tell me near as you can what my share in your estate would be? If I had it now I could make so much more out of it than you are."[83] Struck dumb by his grandson's effrontery, Overholt admired the boy's moxie but refused to advance the money.

Frick resigned from Marcum and Carlisle for a better opportunity at Copper's, a mourning clothes dealer. Unfortunately, typhoid fever struck him with a vengeance, forcing total bed rest at West Overton. Following a slow recuperation, he kept the books for the family distillery.

In 1870, grandfather Abraham Overholt died, gifting his $385,000 estate to his children. Clay's mother, Elizabeth Frick, received $40,000 as her share but twenty-one-year-old Clay nothing. The absence of an inheritance crushed the youth, but he borrowed $10,599 from his mother, about $185,000 in current dollars. With the funds, he and three cousins partnered in a coal mine venture on a 123-acre plot of land adjoining the Broad Ford distillery plus fifty beehive ovens to convert the coal to coke.

Coal came naturally to Frick. Grandpa Overholt had invested in coal mines years earlier. Cousin and partner Abraham Overholt Tintsman owned a piece of the nearby Morgan Mine. Over the past decade, coke had become the fuel of choice for the iron industry, and Frick envisioned its growing future. A thirty-mile by two-and-a-half-mile stretch surrounding the city of Connellsville in Fayette County contained a seam of the country's richest low-sulfur coal buried less than 350 feet below the surface.

Frick had researched the economics of coke prior to investing. Beehive ovens inexpensively baked coal into flat, dense slabs called "coke." The resulting silver-gray metallic carbon slabs provided a far more efficient fuel for Pittsburgh's iron producers than the charcoal previously used. The close access of the mines and ovens to river and rail transportation made shipping easy.

The average coal miner assisted by a teenage apprentice might dig and load ten tons of coal each day and recoup $2.50 for their combined labors. Horse-drawn carts dragged the output to the beehive ovens for conversion to coke. Masons constructed the typical beehive oven, named for its resemblance to a beehive, using roughly 2,500 bricks at an average cost of $300. Most ovens measured five to seven feet in height and eleven to twelve feet in diameter. Each oven lasted for around five years, required minimal maintenance and produced an annual coke output of nearly seven hundred tons.

Blackened muscular laborers shoveled seven to seven and a half tons of coal from carts down a chute atop each oven. A "leveler" evened the piles of coal through a front opening prior to the "burn." A "bricker" cemented the opening with clay to control air input. The oven baked the bituminous coal at 1,600 degrees Fahrenheit for forty-eight to seventy-two hours to burn off sulfur and phosphor impurities. The "puller" inspected the color of the flame, tore down the clay wall at just the right moment and removed the properly cooked cinder, cooling it with a hose. "Lifters" drew the resulting yield of 100 to 165 bushels weighing about four and a half tons and forked it into wheelbarrows for transportation by barge or rail to Pittsburgh's iron foundries.

Flames poured from atop the ovens dotting the hillsides surrounding the mines, oozing a noxious odor while tattooing the nearby rows of workers' houses with smog and grime. The mine operators sold their coke to the iron mills at the going-commodity rate of $1.37 per ton. While the Connellsville laborer struggled with filth, disease, smoke, foul air, injury and penny-ante wages, the bosses raked in the dollars by supplying tens of thousands of tons with minimal cost and investment.

The rush of business consumed the numbers-oriented Frick—mining, production, shipping, sales and profits. He owned plenty of coal land but required additional beehive ovens to bake more coke into dollars. He knew just where to turn for a loan. Frick's grandfather Overholt had been farmer-friendly with Judge Thomas Mellon's father. In fact, Frick's mother, Elizabeth, and the judge had played together as children. Clay contacted the judge for an appointment at his bank, T. Mellon and Sons. Dressed in a conservative dark suit and tie, Frick boarded a train for Pittsburgh.

The fifty-seven-year-old judge had a reputation as a stern, no-nonsense stickler for decorum. The banker valued both time and money as precious commodities. He considered love overrated and appraised marriage strictly by its utilitarian worth. Once he had reached the age of thirty, he searched for a wife of good family with a suitable dowry, capable of providing home-cooked meals and producing suitable heirs to stock the Mellon dynasty. He selected plain Sarah Negley based on his investigation of her "respectable Presbyterian Scottish and Irish stock notable only for good habits and paying their debts."[84] She suffered no health disabilities, possessed rich family connections and appeared of an ideal age, all requirements for the proper Mellon wife.

The judge pursued the wooing process like a business takeover. He accepted Sarah's chattering about scrapbooking and botany, "but to her credit, I must say she never inflicted any music on me," Mellon wrote in his memoirs.[85] "There was no love beforehand so far as I was concerned, nothing but a good opinion of worthy qualities. Had I been rejected, I would have felt neither sad nor depressed, only annoyed at a loss of time."[86] The long-suffering Sarah made an ideal mate, providing children, tending to the house and accepting the strong-willed banker's domineering ways.

The judge and Frick proved kindred spirits in many respects. Mellon listened to Frick's clear, concise proposal to build ovens with the cold, emotionless eyes of a shark while silently approving the bookkeeper's directness. At the conclusion of the presentation, the judge granted a $10,000 loan at 10 percent interest with Frick's father's farm as collateral.

Frick returned to Mellon Bank a few months later seeking an additional $10,000 for fifty more ovens. A Mellon officer summarily rejected the loan based on Frick's youth and inexperience and the newness of the company. The judge interceded. He sent mining engineer James Corey to determine if the coke works warranted an additional loan. Corey issued a succinct report: "Give him the money. Land's good, ovens well built, manager on job all day, keeps books evenings, knows his business down to the ground."[87] Judge Mellon at once became Frick's mentor and chief funding source. The ovens prospered, and Frick repaid the loans.

With increased oven capacity and improved transportation from the newly constructed Mount Pleasant and Broad Railroad, Frick opened a company store that squeezed needed nickels and dimes from the growing force of miners. He price-gouged his cash-strapped workers who needed supplies between pay dates, rationalizing that the convenience of the credit he offered more than compensated for the high price tag. While mine owners like Frick faced a host of problems, they had little time to waste on health and safety concerns for their workforce. If one poor soul became incapacitated, dozens of hungry immigrants lined up for the open spot.

Frick owned two hundred beehive ovens by 1873. His business proceeded according to plan—that is, until the Panic of 1873 toppled a host of financial institutions, including the National Bank of the Commonwealth, Howes and Macy and Fish and Hatch. Even T. Mellon and Sons Bank teetered on the brink of insolvency. Frick described it as a "horrible time."[88] The rapid closure of iron mills stalled coke demand. Thousands of commercial firms closed, but Frick battled the recession with intensified sales efforts. He trekked from plant to plant, hawking coke to anyone who would buy. Long hours and aggressive pricing provided the recipe for survival. He wrote in his notes, "Got up at six, looked over the ovens and set things going, took the train to Pittsburgh at seven, reached the office at ten, legged it from factory to factory soliciting orders 'til three, reached home about six, and attended to the details of mining 'til bedtime."[89] He paid his workers in Frick dollars when necessary, redeemable for basic necessities at the company store.

With business in the doldrums, Frick obtained the rights to the ten-mile Mount Pleasant and Broad Ford rail spur from its cash-strapped investors at bargain basement pricing and brokered it to the B&O Railroad for a $50,000 profit. The funds obtained allowed him to pay off debts and buy out his cousins. Although the scheme was legal, the struggling sellers questioned Frick's morality in taking advantage of them.

The death of grandmother Maria Overholt, the brutal pace of keeping his company afloat through the economic recession and the cumulative aftereffects of childhood illnesses tested Frick's health. His weight dropped, and inflammatory rheumatism racked his body. Pain radiated through his legs, and he walked with difficulty. By 1874, his condition had worsened such that his cousin and former partner Abraham Tintsman wrote in his diary: "Clay is in critical condition." He lacked the strength to wash or dress and hobbled about on crutches for months. He would recover enough during the following year to chase the dollar anew just in time for the country's return to prosperity.

Bad times had presented opportunity for the prepared. In the wake of the depression, while prices were still low, Frick purchased additional coal acreage and beehive ovens. He optioned Morgan Mine stock shares from his nearly bankrupt cousin Abraham Tintsman in October 1875. As business improved, he moved up the rung to the upper echelon of Pittsburgh society. Frick dressed impeccably and moved from Connellsville to Pittsburgh's Monongahela House, a bachelor's tony paradise. His Mennonite grandfather Overholt surely would have lauded his newfound success, if not approving of his fancy outer trappings. The Coal and Coke Exchange elected up-and-comer Frick as its board secretary. He clearly had become a man on the rise. Mellon introduced him to his son Andrew, noting, "That young man has great promise. He will go far unless he over-reaches. That is his only danger."[90] The judge's taciturn son Andrew and Clay would develop a lifetime friendship and business relationship.

Frick worshiped the dollar like most of his fellow industrialists, with slight regard for the immigrants who toiled at his mines and ovens. Whenever human needs vied with profit, Frick chose profit. Steel and ironmongers, coal miners, coke bakers, common laborers and railroaders silently chafed under the yoke of servitude. Poor pay, long hours and dangerous work bred discontent. The slightest economic nudge generated the fodder to thrust the workforce over the edge. A full-scale Pittsburgh railroad wage war erupted in 1877 with volcanic intensity. The brakemen rioted when President Thomas Scott of the Pennsylvania lengthened the number of freight cars per train and instituted reduced pay rates—a double whammy. "The bastards is takin' the vittles from the mouths of our families. They've been robbing us for years. It's time they get a little of what's coming to them," screamed a furious striker. Scott responded with a hard-nosed approach: Give them "a rifle diet for a few days and see how they like that kind of bread." Labor answered with violence.

Mobs destroyed more than one hundred locomotives and burned the Union Depot in Pittsburgh to the ground. Emboldened by whiskey, looters ransacked stores and stole rifles, knives and ammunition. Anarchy ruled. Management blamed outside German agitators and summoned the militia to protect their property. Soldiers attacked, shooting demonstrators who "marched through the streets declaring vengeance and robbing stores where any arms were to be had. Matters grew from bad to worse. Men out of work and who having nothing to lose were only too glad for this opposition," B.F. Jones wrote.

The spirit of rebellion spilled over from the Pennsylvania Railroad to Jones and Laughlin's day workers, who demanded a 25 percent pay increase. The more vocal members urged firemen and foundry workers to join the protest. When management attempted to bring in six scab laborers, the strikers scared off all but two. Jones shuttered the plant in response rather than risk violence. Unrest at Jones and Laughlin simmered through the balance of the summer. Around September 18, hungry workers slowly returned with their tails hanging between their legs. Jones blacklisted the worst troublemakers and jotted in the Jones and Laughlin journal: "The firm has made no concessions."

Frick watched the railroad unrest spill over to his coal fields. When sympathy strikers refused to work, he ousted them from company-owned housing. After one protester refused to leave, the 130-pound Frick, with a deputy's assistance, tossed the offender and his personal contents into a nearby river.

By late fall, the militia had crushed the vandalism on Pittsburgh's streets. The Pennsylvania Railroad had incurred horrible damage from the rampage: 39 buildings, 104 locomotives and 1,245 rail cars had been trashed. The wanton destruction earned labor little sympathy. In addition to twenty deaths, twenty-nine others suffered serious injury, and hundreds bore bumps and abrasions. Management had won the costly war. The thoroughly beaten miners, the railroad brakemen and the Jones and Laughlin crew each begrudgingly returned to work without gaining a single demand. Life returned to normal: the workers worked and the bosses bossed.

• • • • •

Frick now owned three thousand acres of rich Connellsville coal land. He leased the Anchor and Mullen ovens near Mount Pleasant so his coke output could match his coal input. In 1879, he acquired a coal-rich farm along the South West Pennsylvania Railroad, which he organized into the Morehead Coke Company. With employment exceeding one thousand and 80 percent control of the region's coking coal, Henry Clay Frick deservedly earned the title of the "Coke King." When coke pricing vaulted to $5 per ton, he pocketed nearly $20,000 daily. While miners hacked and coughed, trees died from poisoned smoke and sinkholes blighted the landscape, Frick counted his coins.

December 19, 1879, presented a watershed moment. Clay finished a game of checkers with his cousin on the porch of the very Mount Pleasant store where he had clerked a decade earlier. He gently pushed back the checkerboard and lit a cigar. Exhaling ever so slowly, he rehashed the challenges he had overcome: the inheritance never received, rheumatic fever, rheumatoid arthritis, the death of his beloved grandparents, physical pain and the stress of business. A waft of smoke circled above his head like a coven of ghosts haunting his thoughts. He had risen above it all. Life had sharpened him like the blade of a bowie knife, desensitized to the daily struggles of the workers who burrowed beneath the bowels of the earth in search of coal or sweated by the flames of his beehive ovens. Cave-ins, explosions and sickness were a cost of doing business. Like a commander marching his troops to the parade field, he watched with detachment as his workforce disappeared from sight. Victory demanded toughness. If his work required predatory skills, he determined to be the fiercest predator in the jungle.

He sucked in another drag from his five-cent stogie and exhaled effusively. As the smoke ring took shape hovering above his head, he mentally tabulated his financial worth: "Morewood's 500 beehive ovens, Connellsville's 853, the coal acreage and let me see." Frick took another drag from his smoldering cigar and smugly congratulated himself. He had become lord of the coal fields, the king of coke. By his reckoning—and he rarely made mathematical errors—he counted himself a millionaire.

• • • • •

While those who labored in the furnaces at J. Edgar Thomson, Homestead and Duquesne struggled with heat, dirt and danger, the coal miners suffered even more. An experienced miner grossed one dollar per ton during boom periods, earning up to six dollars per day, a significant income, but nuisance fees for tool sharpening, lamp oil and blasting powder reduced take-home pay back to subsistence levels. Economic troughs frequently cut hours and wages even further. Overcharges at the company store between paydays for basic necessities like flour and sugar added to the workingman's woes. Miners toiled in darkness, inhaled poisoned air, lived in substandard housing and ate unhealthy foods, but America offered few decent jobs for its newcomers.

In contrast, Frick relished the trappings of success—the fancy clothing, the exclusive clubs and, most of all, the respect of his contemporaries. He toyed with the idea of running for Congress, but Judge Mellon dissuaded him: "You can do so much more as a private businessman without the incursions of others." He joined King Solomon Lodge Number 346 of the Free and Accepted Masons of Connellsville, serving as secretary-treasurer. The York Rite granted him high honors, including the Order of the Knights of Malta and investiture as a Knight Templar. Benjamin Ruff, a fellow Monongahela House resident and coke dealer, asked him to join the newly formed South Fork Fishing and Hunting Club on Lake Conemaugh, a few miles from Johnstown. The invitation delighted Frick, barely thirty years of age, who joined a club that would include Henry Phipps, Andrew Carnegie, Andrew Mellon, Philander Knox, John Leishman, James Reed and Robert Pitcairn among its elite membership of only fifty. The club provided hunting, fishing, boating, riding and hiking paths for Pittsburgh's rich. The lake's leaky dam created a nuisance, but the membership jury-rigged it with mud and straw as a temporary stopgap.

Flush with coke cash, Frick invited twenty-five-year-old Andrew Mellon, the recently promoted president of T. Mellon and Sons Bank and a fellow South Fork member, to join him for an European holiday. If contemporaries considered Frick tight lipped, Mellon spoke even less. The frivolous offer seemed so unlike Frick, the acceptance so unlike Mellon. The night before their sailing from New York, the bachelors meandered along Fifth Avenue, strolling past the mansions on millionaires' row along with other gawkers, puffing on cigars like they owned the world. The magnificent estates sparkled like fairy tale castles under the illumination of gaslights. Pointing to the expansive Cornelius Vanderbilt II palace on Fifty-Eighth Street, Frick asked Mellon for a guess on the upkeep of such a monster. Mellon answered, "I'd say $1,000 per day or north of $300,000 per year." Frick said, "That is all I shall ever want."[91] One day, Frick would rent that very mansion.

The trip cemented Frick and Mellon's friendship, as well as their mutual love for art, a gift all America would one day enjoy. Shortly after his return to Pittsburgh, Clay purchased his first major work of art, George Hetzel's *Landscape with a River*, earning Mellon's hearty approval.

A few weeks later, the bachelor pals attended a gala where a beautiful woman sashayed past them. The girl's ocean-blue eyes, rich brown hair, exquisite thin lips and angelic smile mesmerized Frick. "There is the handsomest girl in the room. Do you know who she is?" Mellon recognized Adelaide Childs as the daughter of Asa Childs, a shoe distributor who sold boots to Frick's company store and a cousin to Henry Phipps's wife, Anne Childs Shaffer. Frick implored Mellon for an introduction. Mellon asked an older man he knew for assistance, according to proper etiquette. Desire overcame Frick's customary reticence. With the *savior faire* of a lovesick puppy, he approached the twenty-two-year-old beauty on his own and announced with a slight bow: "Good evening, I am Clay Frick." The young lady smiled. After several minutes of polite chitchat, Adelaide Childs granted Clay permission to call on her the following Sunday.

Lovely Adelaide captivated Frick. Within a few short months, he proposed, and she accepted. The wedding took place in Pittsburgh on December 15, 1881, with Andrew Mellon as best man. Tom Carnegie tipped off his brother that the couple would be honeymooning in New York. Andrew Carnegie invited Clay and Adelaide to lunch with him and his mother at their hotel quarters. Carnegie proposed another marriage: the merger of H.C. Frick and Carnegie Steel. Clay and Adelaide remained a loving couple through a life of tribulations. Carnegie and Frick's affair would end in a bitter divorce.

Frick purchased a $25,000 two-story, eleven-room, Italian-style home at the east end of Pittsburgh from Carnegie partner John Vandevort as a belated wedding gift for his new bride. He would call the house Clayton, a derivation of his middle name. The couple budgeted an additional $50,000 for its enlargement into a twenty-four-room Victorian château.

Adelaide and Clay Frick decorated the interior with French Empire and Chippendale antique furniture from New York, and they dressed the walls of the music room in velvet and the drawing room in rose brocade. Artisans etched Frick's initials "HCF" into the breakfast room's massive mahogany chairs. Elaborate pewter fixtures and a massive chandelier dignified the entryway. English Copeland china and a kettledrum tea service adorned the dining room table. As a devotee of the latest in technology, Frick installed one of Pittsburgh' s first telephone systems in his office and allowed his friend George Westinghouse to electrify the house.

Henry Clay Frick's home, Clayton.

Although grandfather Overholt had considered cards the work of the devil, Clay matched whist or poker wits many an evening with R.B. and Andrew Mellon, George Westinghouse, Philander Knox and occasional visitors like Nicola Tesla. Although he concentrated on coke manufacturing six days per week, he set aside time for family on Sundays and dinners. During the day, Adelaide, who enjoyed the society of the wives of George Westinghouse, John Pitcairn, George Dilworth and Tom Moorhead, oversaw a staff of eleven, which included a cook, five maids, two stable hands and a groundskeeper. At dinner, Adelaide summoned her maids for each course with the gentle tinkle of the bell that sat by her place setting. On March 12, 1883, Adelaide gave birth to her first of four children, Childs, a boy. At this moment in time, life was magnificent for Adelaide and Clay in their dream mansion, Clayton.

• • • • •

Immigrant mill families had to watch every penny, always only one mishap away from the poorhouse. The lucky ones managed to squirrel away enough for a tiny row house on an unpaved street, nestled among rows of similar soot-covered houses—all within walking distance to the plant. A hand pump might deliver running water in the kitchen, an outhouse in the tiny backyard providing the necessary convenience. For them, luxury was sitting by the light of the brick fireplace in an easy chair. If the house needed paint and repair, the worker undertook such chores himself, provided he could come up with a few coins for the supplies.

Ghetto neighborhoods contained a mishmash of Czech, Hungarian, Serbian, Russian, Ukrainian and Polish immigrants. Women in babushkas and aprons spoke little or no English and chattered in a cacophony of languages unintelligible to outsiders. The ladies wore dark colors since smoke and dirt stained lighter shades black. Most of the homes had been painted white, but age and pollution had stained them gray. The men went by comfortable old-world names like Stosh, Stanislaw and Lech. The sweet smell of home-baked pierogis, uszka dumplings, cheese blintzes, hybivka mushroom soup and fried dough called pampushky masked the surrounding stink from the mills.

•••••

While Adelaide managed the house, Clay Frick ruled his commercial domain with iron resolve. He demanded autonomy to run his coal empire the way he saw fit, but gadfly Andrew Carnegie constantly interfered. The two entrepreneurs clashed, and although both spoke the same language, neither understood the other. When Carnegie vetoed a proposed Standard and Trotter Coke acquisition, Frick simmered, petulantly responding, "I do not like the tone of your letter in the matter of the properties in question. I shall have to differ from you, and I think the future will bear me out."[92] Carnegie eventually caved, allowing Frick his way.

In April 1884, Frick and Carnegie bickered over coke pricing. The Connellsville Coke Syndicate proposed an overall rising of rates to $1.20 per ton. Senior partner Carnegie, fearful of the negative impact on steel costs, undermined Frick and vetoed the agreement. Frick fumed but held his tongue. Frick carped, "I am pleased to possess two ears and just one mouth. Mr. Carnegie seems to have many mouths and no ears."

•••••

B.F. Jones watched Andy Carnegie dominate the steel market, leaving Jones and Laughlin in the dust. Jones intended to shift from iron to steel manufacturing in the imminent future, but politics absorbed him in 1884. The National Committee of the Republican Convention appointed him chairman despite his lack of political experience. B.F. addressed his co-committee members on June 2:

> *Gentlemen of the National Committee, I did not seek nor did I expect this distinguished honor. I accept your action not as a compliment to myself personally, but as a recognition of our great business interests. In accepting this important position, I have many misgivings as to my ability to perform the duties involved satisfactorily, and only do so with the understanding that the other members of the committee will not only assist, but give the full benefit of their superior experience, judgment, and energy in conducting the campaign. Victory will be ours in November as surely as the sun will bless us with its light.*[93]

The Chicago Republican Convention nominated Senator James Blaine to oppose Governor Grover Cleveland of New York for president. For the next four months, Jones lived and breathed the presidential campaign from the Windsor Hotel in New York. The Republican Party proposed a high-tariff platform, which Jones supported. In one article, he wrote:

> *I am a protectionist because our country has prospered with protection and languished without it. Because protection provides diversified employment and largely relieves wage earners from foreign competition, thereby enabling them to be liberal consumers as well, because the theory of free trade between nations is as fallacious, impractical, and utterly absurd as that of free love.*[94]

A heavy dose of protectionism paved the way for iron and steel's steady growth. With business booming, skilled workers earned enough pay to splurge on new dresses for their wives, toys for their children and perhaps even an easy chair for the living room.

A host of accusations and counter-accusations marred the presidential campaign. The media accused Cleveland of siring a bastard child and Blaine of taking a bribe. The election outcome teetered from side to side until Reverend Samuel Burchard in New York labeled the Democrats "the party of rum, Romanism and rebellion," generating a Catholic tidal wave of voter resentment and securing Grover Cleveland a slim margin of victory.

Despite the loss, the Republican leadership lauded Jones for a job well done. Few blamed him for the Blaine defeat. Colonel Stephen Elkins wrote in the November 19, 1884 issue of the *Maitland Express*: "Mr. Jones proved himself an able, faithful and loyal chairman. He showed great executive ability, is a self-made man of great wealth, and thoroughly understands the politics of the country. In my judgment, he was as good a chairman as the National Committee ever had." The December 26 issue of the *New York Daily Tribune* praised Jones as "most sagacious, experienced and earnest, broad in his views, commanding the respect and confidence of all who come in contact with him."

With the national election over, the American Iron and Steel Association elected Jones as president to replace retiring Cambria Steel CEO Daniel Morrell. Jones wholly endorsed the organization's motto: "Protection to home industry." The *Philadelphia Times* announced, "The Iron and Steel Association has paid a fitting compliment to one who has labored long and effectively in the cause of protection. At the same time, the acquisition of a man so widely known and of so much influence cannot fail but add strength to the association." Jones would hold the presidency for the next eighteen years, also serving as president of the Western Iron Association.

Labor considered Jones tough but fair. When Amalgamated Union president William Weihe asked for a half day off in the middle of the week for an industry picnic holiday in the mid-1880s, a highly unusual request for the times, Jones consented. The Carnegie and Oliver companies followed suit, and thousands attended an afternoon of games and speeches. Crowds of employees and their families enjoyed fried chicken, beans and pie. The children played tag and kick the can and ran relay races.

Although Carnegie led the country in steel production, Jones aimed for a larger slice of the action for Jones and Laughlin, investing in two seven-ton Bessemer converters. Ratcheting up the ante, his company constructed the city's first steam-powered, steel-bar blooming mill for the reduction of giant ingots into smaller billets. He also built fifteen high-temperature soaking pits to ensure metal uniformity. Jones and Laughlin's first Bessemer heat on August 19, 1886, pushed the company into the steel age. In typical no-nonsense fashion, Jones ordered the immediate construction of a third Eliza blast furnace to ensure an adequate supply of pig iron to feed his converters and contracted with a Cincinnati firm for the erection of a new sheet and roller mill to shape the steel bars into finished products. When Jones learned two contractors accidentally drowned while on the job, he displayed little sympathy once he learned they had been drinking. Rather, he fumed at their sinful excess and the needless delay it caused for his grand plan.

In a further advancement, Jones and Laughlin built a 321-foot metal railroad bridge spanning the Monongahela River to link the steel manufacturing plant with the Eliza furnaces, replacing an unwieldy barge transportation system. Rail cars teeming with molten pig iron crossed the bridge and dumped their loads into the Bessemer converters. A separate internal rail system carried giant steel ingots to the blooming and slabbing mills for the production of billets or bars. Rail cars and barges delivered the finished steel product to the Pittsburgh distribution center for eastern customers or to the Chicago or St. Louis distribution centers for western users. After decades of concentration on iron products, Jones and Laughlin had developed into a competitive force in steel and a threat to the Carnegie Empire's domination.

• • • • •

With the depletion in Carnegie leadership from death, retirement and firings, Carnegie promoted Frick as president of the steel company on January 31, 1887. The initial weeks went well, with Frick overseeing both the coal and the steel companies. Carnegie wrote to his second in command: "Take good care of that head of yours; it is wanted." The board sweetened Frick's financial package with a 2 percent stock-ownership position in the steel company, paid through future dividends. To seal the arrangement, Frick signed the Iron Clad Agreement previously drawn up by Phipps, which required a mandatory company stock buy-back at book value in the event of his retirement or involuntary separation. Little did Frick realize the impact that the Iron Clad would play in his future.

• • • • •

The Mine Laborers Amalgamated Association had ramped up its organizational efforts in the Connellsville region during 1887. Poor working conditions, low pay and long hours forced the miners to face the specter of injury every day. Frick had invested hundreds of thousands of dollars purchasing coal land, ovens and equipment. Every fiber in his body bristled at the growth of union interference in his affairs. Frick banded together with the Coal Syndicate in unified opposition to organized labor's demands. When union negotiators pursued a ten-cent-per-ton pay increase, the syndicate refused. A defiant union struck on May 7. Frick responded with an expulsion of the strikers from all company-owned housing.

Violence and coal went hand in hand—unfeeling management battling an army of hungry, tough laborers, many of whom spoke little English. Fearful of damage to company property from the unruly rank and file, Frick hired 150 armed Pinkerton guards to protect his property.

The American appetite for steel had multiplied over the past few decades. In a growing industrialized nation, the need for rails, naval armor plate, structural skyscraper skeletons, saws, nails, shovels, hammers, axes and farm implements had soared. The coal strike had shut down seven Carnegie mills. Frick had warned the company to maintain backup coke inventory as a cautionary measure, but the purchasing agents failed to follow through on the recommendation.

Carnegie liked to portray himself in his writings as an enlightened employer. His essay "An Employer's View of the Labor Question" espoused the union's right to organize, but he lacked the human relations insight of Henry Heinz or George Westinghouse. Carnegie's hypocritical drivel infuriated Frick, who avoided contact with the media unless absolutely necessary. Although Carnegie's words told a tale of beneficence, his actions revealed the soul of a labor exploiter, just like Frick, who viewed his miners as mere production tools. He advocated paying just enough for food, clothing and housing—the basics—anything more would be a waste of resources. When labor, like disobedient children, rebelled against his authority, Frick chafed.

For Carnegie, the bottom line trumped all. With the coal strike squeezing fuel supplies, the flow of steel profits slowed in the company coffers. Carnegie refused to accept a disruption. He cabled his lieutenant: "No stoppage tolerated!" This short-range capitulation maintained the cash flow but undermined his chief operating officer.[95] Frick fumed at Carnegie's lack of guts to stare down a half-baked union ruckus, forcing him to abandon his Coke Syndicate brothers. He fired off a note in disgust: "I object to so manifest a prostitution of the coke company's interests in order to promote your steel business." Carnegie snidely replied that Frick might elect to spend more time at the coke company.

An embarrassed Frick settled with the union as ordered for a seven-cent-per-hour increase. The other syndicate members held firm until the defeated strikers returned to work at their current rate after just a few weeks. The forced settlement placed the Carnegie plants at a coke price disadvantage against the competition during the balance of the contract. Frick resigned from the Carnegie presidency in disgust and departed Pittsburgh for an extended European family holiday. Henry Phipps reluctantly stepped back in as a temporary replacement.

Phipps and George Lauder pleaded with Carnegie to swallow his pride and apologize for sabotaging Frick. Since Carnegie rarely held a grudge, he invited Frick to his rented home in Scotland to hash out the issues. Frick proved a harder case. He rarely forgave and never forgot, but Phipps convinced him to sit down with Carnegie.

During the powwow in Scotland, Frick and his wife received an unnerving telegram from their governess in London. Their baby daughter Martha, whom they lovingly nicknamed "Rosebud," had lost her appetite and developed a limp. Unbeknownst to the Frick family, she had swallowed a pin. Adelaide and Clay Frick sped to their daughter's side. When Martha's health improved, Frick returned to Scotland to complete the reconciliation. The November 10, 1887 *Pennsylvania Press* headline announced: "Frick Is Boss Again!" Carnegie anted $90,000 to the Coke Syndicate coffers as compensation for his interference and promised Frick future autonomy, a pledge so contrary to his meddlesome personality. Again at the helm, Frick intended to execute a high-growth strategy grounded on staunch anti-unionism and unbridled management control.

• • • • •

The *Oil City Derick* floated the name of B.F. Jones as a potential candidate in the 1888 presidential race, but the elder steel statesman opted to concentrate on his steel business and Pittsburgh community interests. Directorships in banks, coal and coke companies, iron mines, insurance concerns, the Duquesne Club and nonprofits filled his time. He served on the committee for the Allegheny Cemetery incorporation and the 1888 Pittsburgh Centennial Committee along with Andrew Carnegie and Judge Thomas Mellon.

Jones authored the article "Protection" in the *North American Review*, advocating tariffs to protect the American workingman. *Iron Age*, the industry newspaper, dubbed him the "most respected man in the iron trade," and Carnegie referred to him as "Nestor," the sage Athenian described by Homer in the *Iliad.* Labor joined Jones in its support of the tariff as a defense against foreign competition. When Jones traveled to Washington to lobby, the hotel doorman greeted him with a snicker: "I think I smell a tariff coming." Pittsburgh limerick writer Arthur Burgoyne joked about Jones: "But in case the loss threatens to hurt his affairs, he gets congress to run up the tax on his wares. Then he bids all his subjects to raise a hurrah for the blessings conferred by the new tariff law."[96]

On April 14, 1888, Miles Humphreys, the Puddlers Union president, signed a contract employing the sliding scale as designed by B.F. Jones. When steel prices rose, wage rates would increase. When iron prices fell, wage rates dropped. Since the puddlers earned a higher per-ton rate during the present good times, they willingly inked the agreement. Problems would only arise when prices declined and management sought to enforce reduced wage rates.

April 1889 delivered joy to the B.F. Jones family. Daughter Alice married iron manufacturer William Willock at the Jones Irwin Street home. Following the ceremony, the Duquesne Club sparked with a gala reception featuring a who's who of Pittsburgh society. A week later, tragedy struck. Brother Thomas Mifflin Jones unexpectedly died—an omen of other disasters to follow.

• • • • •

The South Fork Hunting and Fishing Club served as the summer playground for Pittsburgh's steel titans. Engineers had cautioned the club's leadership about the weakness of the dirt dam, the largest such structure in the country, which held Lake Connemaugh in check. The club leadership ignored the warning. When six inches of rain overfilled the lake during a twenty-four-hour May 31, 1889 storm, a torrent of water exploded through the weakened dam and crushed the nearby Woodvale's Gautier Ironworks with triple the lethal force of Niagara Falls.

Within minutes, a flotsam and jetsam of barbed wire from the ironworks, broken railroad tracks and shattered beams poured a sixty-foot tsunami

Destruction from the Johnstown flood.

of death on the unprepared city of Johnstown a few miles distant. Oil tanks ruptured and burst into flames. Hundreds who survived the original onslaught drowned in the four-square-mile quagmire left in the flood's wake. All in all, 2,209 inhabitants perished, and Johnstown burned for three days.

Upon viewing the carnage, Robert Pitcairn of the Pennsylvania Railroad assembled good Samaritans from Pittsburgh to assist in the cleanup. Captain Jones organized a J. Edgar Thomson volunteer crew, barking at his men: "Will any of you boys join me? There ain't no pay, but you'll be doing a fine deed." Dozens volunteered. The horror of the devastation assaulted the men. This was the devil's work. The flood had demolished 1,600 homes and 280 businesses, slaughtering innocent women and children. Water covered the entire downtown, which stunk like an open sewer. A dead dog floated past, but that was not the worst. Grown men gagged at a handful of corpses, whose empty eyes stared vacantly into space, their mouths pleading silently for help.

A volunteer doctor ordered an immediate mass burial to stem the spread of disease. The men understood. The bodies reeked from the stale aroma of death. Flies swarmed about the decomposing heads. Sorry-faced automatons dropped the rotting corpses inside pine coffins and stacked them on wagons. A baby, an old lady wearing an apron, an infant wrapped in a blanket, a fat man with no shirt, a cadaverous old-timer with a gash across his mouth and barbed wire slicing through his waist, some fully dressed, others clad only in nightshirts or undergarments—all were carted for burial to the Plot of the Unknown at Grandview Cemetery. The corpses of 777 victims never would be identified.

One Johnstown survivor grumbled, "How could God allow this disaster to take place?" The world knew. The South Fork Hunting and Fishing Club members had ignored repairs on the dam, opting to fish and boat rather than take responsibility for the safety of others—just like they did in their own businesses. The headline of the June 7 issue of the *New York World* declared, "The Club Is Guilty!" Unfeeling titans like Frick, Phipps and Mellon received the lion's share of blame. Poet Isaac Reed penned the couplet: "All the horrors that hell could wish, such was the price that was paid for fish."[97]

Fearful of a lawsuit, the club membership called an emergency meeting and adopted a policy of public silence. Henry Phipps and Henry Clay Frick appointed a five-member relief committee that campaigned to raise funds for temporary housing and food. Andrew Carnegie wrote his partners: "That South Fork calamity has driven all else out of our thoughts."[98] Carnegie Brothers, Carnegie, Phipps and Company and H.C. Frick each donated $5,000 to soothe their corporate consciences. Andrew Mellon sent $2,000 and Benjamin Thaw $3,000. Although nearly $3 million flowed into

Johnstown and the club closed forever, accusations would follow members throughout the remainder of their lives.

Once the adverse publicity from the South Fork tragedy dropped off the front pages, Frick concentrated on business. In a brilliant move, he acquired arch-competitor Duquesne Steelworks for Carnegie in October 1890 for a song. In what some experts called "the greatest bargain in the century," the purchase garnered the patented process of running hot ingots through the rolling mills without reheating.[99] Frick also bought the J.A. Strickler coal lands and an interest in the Hostetter-Connellsvile Coke Company to round out H.C. Frick's extensive portfolio.

Despite their differences, Carnegie raved about his partner: "Frick is a marvel! Let's get all F's."[100] Clay reached his desk early in the morning and rarely left before six o'clock. He vacationed infrequently, focusing on efficiency and productivity—molding the decentralized Carnegie holdings into a sleek profit-making machine. Operating like a closed-entry accounting system, he rarely trusted others with his full confidence. Yet he ingratiated himself with important politicians ranging from United States ambassador to Russia Andrew White to Senator Boies Penrose. The Frick-Carnegie partnership had reached a high point. Only health worries about daughter Martha gnawed at Frick.

• • • • •

Labor unrest mounted in western Pennsylvania during the last decade of the nineteenth century. At J. Edgar Thomson, the tension of twelve-hour days and miserable working conditions strained the patience of the workforce. The men were tired and overwrought. In January 1891, Frick and Superintendent Charles Schwab suppressed a riot at the plant, which they termed "a drunken Hungarian spree."[101] Pinkerton enforcers carried the day for Frick, but abuse had primed the workforce like a bomb about to explode. A Polish worker complained to his coworkers at the saloon: "We are mules for hauling. When we drop they fill our spots with new mules for pennies. Only union can protect us. We need union." Others nodded but said nothing. They needed their jobs.

On January 27, 1891, a gas explosion at Frick's Mammoth Mine Number 1 snuffed out the lives of 107 miners. The intensity of the burst strewed piles of charred masses of flesh one hundred feet under ground. Volunteers buried the mishmash of nearly 80 unidentifiable bodies into a common trench at St. John's Cemetery in Mount Pleasant. Michael Lukosh, John Sauteteske, Martin Shavinsky and Andrew Charensky were among the host of unfortunate Eastern European immigrants who lost their lives. Many families were left penniless, but to the company's credit, H.C. Frick Coke provided some cash assistance.

The rash of explosions and injuries caused politicians and labor alike to clamor for safety measures. On February 9, the newly formed United Mineworkers flexed their muscle: ten thousand workers walked off the job in an organized protest. The men demanded better protection from injury and higher wages. Frick countered with a 10 percent decrease in wages rather than the requested increase. He had stockpiled coke in preparation for the worst. H.C. Frick's working conditions and pay matched the industry norm. He refused to offer more than the competition.

As the miners suffered through the strike, tempers snapped. A mob of 1,500 rioted on March 30. The coal fields exploded with violence. A bomb blast echoed across the West Lessening Works. A handful of thugs grabbed Superintendent Morris Ramsay and bloodied his nose. The sheriff threw 3 of the offending unionists in the Greensburg jail for assault. On April 2, nearly 1,000 militants marched to the accompaniment of band music, heavily armed with revolvers, iron crowbars, bats and knives. The mob destroyed telephone and telegraph lines leading to the Morehead Works, burned wheelbarrows, smashed tools and refused to disperse. The Moreland defenders panicked. Shots rang out, and 43 men fell, 6 dead on the spot: Paul Donahas, Valentine Zeidel, Josef Brochto-Procte, Jacob Shucaskey, John Fudora and Anda Rist, each felled by a single lethal shot. Within days, 3 more would die from their wounds.

Frick threatened further violence with more violence, promising, "The fight from this time will be bitter."[102] On April 4, management summoned the National Guard. Two regiments of soldiers shielded company property. Repulsed by the armed troops, the strikers sulked away to bury their dead. Frick gloated, "The scoundrels have no one to blame for it but themselves."[103] By May 25, the beaten workers had returned to their shovels and ovens under the company's draconian terms. The failed strike at Morewood set back the United Mineworkers organizing efforts for years.

Although Frick displayed minimal compassion for his immigrant drudges, he did act with generosity toward the members of his management team. When Morehead manager Ramsay died from pancreatic cancer at age forty-four, Frick provided one year's salary for his widow and promoted his twenty-two-year-old son to the superintendent's position.

The health of Frick's daughter Martha continued in rapid decline. During the summer, Carnegie opened his Cresson summer home to the Frick family in hopes the mountain air and cool waters might benefit the near skeletal Martha. Carnegie dispatched his personal doctor, Joseph Garmany, who realized little could be done. Frick helplessly watched his daughter fade. His poor "Rosebud" died a painful death on July 28 with her parents by her bedside. A train carried the disconsolate family members to Pittsburgh for the funeral. Hard-hearted Frick had loved his daughter with all his being.

Martha's death tore the Frick family asunder. Daughter Helen asked, "Where is my Tissie?" The nanny told the child she had gone to heaven. "You go to heaven and get my Tissie," Helen implored.[104] The loss of Martha erased Adelaide's smile forever and dented the forty-two-year-old Frick's invincibility. Frick sought solace in work. Concerned with his partner's mental and physical health, Carnegie urged Frick to slow his pace and take a European holiday. Frick ignored the advice and grieved in silence.

Scarcely a day passed at the Homestead plant without some disruption. With the current union contact scheduled to end in July, Frick put forth a tough take-it-or-leave-it offer. Modern steelmaking improvements had skewed labor relations power to the side of management. New equipment had reduced the need for many skilled jobs. Automatic levers, control tables and hydraulic lifts streamlined intra-plant transportation and eliminated a host of other positions. The plant had become a technological wonder with sixteen open-hearth furnaces, a Bessemer converter, an armor plate line, a blooming mill and three structural mills. Each operation incorporated state-of-the-art labor-saving machinery, including a variety of hydraulic, gas, water and coke power supplies. Eighteen "dinky" locomotives and four switching engines facilitated material flows, and the Pittsburgh McKeesport and Youghiogheny Railroad afforded a new transportation line from Connellsville's coal and coke fields to the plant. The massive facility contained fifty acres under roof, and industry experts valued it at a whopping $6 million (approximately $162 million in current dollars). Only the increase in production output prevented massive layoffs.

The Amalgamated Association of Iron and Steel Workers had only eight hundred members and eight lodges in Homestead in 1892. The union represented the skilled trades, the aristocracy of labor, but the unskilled workers who hoped one day to advance up the ladder added their support. The skilled crafts, including the journeymen, first helpers, puddlers, master melters and foremen, might earn as much as $1.48 per hour versus just $0.14 for the common laborer. With the union's ouster at J. Edgar Thomson several years earlier, Homestead remained Carnegie's only organized plant.

•••••

B.F. Jones had grown rich, and he lived a life befitting his success. His Brighton Road, Allegheny City manor stood as one of the larger homes in the city, with eleven servants housed in the adjacent two-story brick carriage house and main building, ready to attend to his every need. Jones and Laughlin was the second-largest steel manufacturer in Pittsburgh. Only the giant Carnegie Works surpassed Jones's company in size. His only son, Princeton-educated B.F. Jr., and nephew William Larimer Jones recently had joined the firm as trainees to add management depth and ensure continuity. The company employed more than three thousand and paid out more than $2,225,000 in wages, with average pay of $12 per week. Annual capacity approximated "350,000 tons of steel billets and blooms, 50,000 tons of muck bar, and 450,000 tons of finished materials."[105] Jones and Laughlin also ran two foundries with a manufacturing capacity of about 85,000 tons in various structural materials.

With the recent implementation of new steelmaking facilities and modern technology, Jones and Laughlin shut down thirty-five puddling furnaces. The puddlers, obsolete relics of an earlier era, had outlived their usefulness, just as they had at Homestead and J. Edgar Thomson, allowing for a tougher stance against labor. Although Jones and Laughlin's own union contract with the AAISW neared its end, Jones sat back to see how events would play out at Homestead prior to the initiation of negotiations.

•••••

President Frick sat in the catbird's seat at Carnegie, Phipps and Company. He owned nearly 6 percent of the company's stock in addition to his shares in H.C. Frick Coal's acreage and 8,050 beehive ovens. Coke profits in 1890 had edged toward $2 million, allowing Frick to pocket $420,367 and Carnegie $498,333 in dividends.[106] To maximize profits at Carnegie, Frick had squeezed every ounce of productivity from his workforce. During Tuesday and Saturday morning supervisory meetings, he questioned the smallest variations in costs and productivity. Advanced technology combined with Frick's push for high output and sales targets generated record Carnegie earnings of $3,540,000 for 1891. Frick now set his sights on reducing labor rates at the unionized Homestead plant to bring about further gains.

Company secretary Francis Lovejoy best summed up Carnegie and Frick's revised labor strategy: the union "seemed to run our mill. We are going to do that ourselves thereafter."[107] Frick and Carnegie contracted former AAISW union leader William Martin to draft new wage scales based on the averages paid by other mills in the region. Employing Martin's findings, the negotiating team of Francis Lovejoy, John Potter and Henry Clay Frick offered AAISW negotiator William Roberts a wage-rate reduction from twenty-five dollars to twenty-two dollars per ton at the January 1892 meeting. Frick intended to bully the union into submission or kill it.

The rout of the United Mineworkers at Morehead had further hardened Frick's heart toward unions. For years, Homestead had been a hotbed of labor unrest. Frick demanded the prerogative to hire and fire whom he chose and to control hours, working conditions and production. As he saw it, Carnegie shareholders had funded an automated plant via the reinvestment of dividends rather than through worker contributions. Capital investment at Homestead paid for the expensive equipment that produced parts for bridges, structural beams and armor plate for warships. Such logic led to the obvious conclusion that the bulk of the financial rewards belonged in management's wallet. Frick thought union wage and working condition demands bordered on extortion. Frick excluded the blood, the sweat, the broken limbs and the lost lives of the steelworkers from the equation. He warned plant superintendent Potter, "If the men do not sign the company scale by the 24th, the Homestead Steel Works will never operate as a union mill again."[108] Union vice-president Roberts questioned Potter on the fairness of the proposal. The superintendent merely shook his head and replied, "I cannot help it. It is Mr. Frick's ultimatum."[109]

Frick had thrown down the gauntlet for a duel with Homestead's 3,800-member workforce. Superintendent Potter rolled reserve naval plate in anticipation of a potential strike. The company stuck with a 15 percent reduction in pay for its skilled workers. Hugh O'Donnell, a former newspaperman and heater in the plate mill, represented the AAISW. He countered with a wage-rate reduction offer of twenty-four dollars per ton. The company softened to twenty-three dollars as a last and final offer, refusing to budge further. A contract termination date of December 31 rather than July 1, a slow production time for the company and the onset of winter for the workers, acted as an additional irritant. In a final slap at labor, Frick demanded the right to reopen the contract should additional technological improvements occur. The labor negotiation spokesman balked, unwilling to take the offer to the men. Frick ended further negotiations. Both labor and management braced for a fight.

Rumors sped through J. Edgar Thomson concerning the labor struggle at Homestead. The laborers had watched their workday expand from ten to twelve hours and pay stall following the union ouster. Should the

Left and opposite: Illustrations of the strike at the Homestead steel plant in 1892 that quickly escalated to violence.

AAISW fall, life at the mill would worsen. As some of the workers took the long way home past Homestead, less than four miles out of their way via the tram bridge, they gawked at the freshly constructed fence surrounding the plant. Three strands of barbed wire guarded the pointy fence tops. Two hundred holes pierced the slats, each just large enough for a rifle barrel. A fresh coat of white paint masked "Fort Frick's" ultimate purpose.

On Saturday, June 25, the company posted notices through town proclaiming the Homestead plant nonunion. Disenfranchised activists tore down the signage while mouthing obscenities. A crowd gathered around straw effigies labeled "Frick" and "Potter," which hung from ropes and swayed to and fro atop an oak tree. Trouble was in the wind. The *Dispatch* headline blared: "Preparing for War."

Attorney Philander Knox hired three hundred armed guards from Robert Pinkerton of Pinkerton National Detective Agency as protection for company property. Plant superintendent John Potter posted a notice on the fence formalizing the policy of dealing with its workforce only on an "individual" basis. He refused to cave like his predecessor William Abbott had done three years earlier. He would drive the union out of Homestead.

On June 28, management shut down the 119-inch plate mill and one furnace. The next day, Homestead closed the remaining departments, locking out almost four thousand workers. As a final thrust, the company announced the reopening as a nonunion plant after the upcoming holiday.

On July 2, the union negotiating team huddled around a conference table at Bosh's Hotel to construct a plan of action. A unionist grumbled, "They closed the damn mill a day before the contract even ended." O'Donnell cautioned vigilance and preparedness: "We got to keep our eyes open and know what they're doing. I wouldn't put it past that bastard Frick to bring in scabs." A man in the back of the room suggested a scout patrol of skiffs along the Monongahela River. O'Donnell seconded the idea. "This is our plant—our bread and butter. Force is all they will respect." Another committee member pounded his fist on the table with visible effect and proclaimed, "We must protect our families and our jobs at all costs." O'Donnell had difficulty restraining the more radical of his cohorts who threatened to shoot any son-of-a-bitch "black-sheep" intruder.

Homestead's churches filled on Sunday, July 3. Priests and ministers prayed for a fair and swift outcome. On the fifth, Allegheny sheriff William McCreary posted "No Congregating" signs outside the plant, but the pro-labor citizenry escorted the outnumbered deputies out of town and tore down the signs.

Furious at the opposition to his authority, McCreary barged into AAISW headquarters at the Bosh Hotel accompanied by two deputies and glared at O'Donnell. He pointed a menacing finger at each man around the table: "I hold you and the rest of your bunch individually responsible for any and all damage to Homestead's plant or personnel. If I have to, I will send a team of armed deputies, and you won't like the result one bit." O'Donnell suggested rounding up a militia of five hundred Homestead citizens in lieu of outside forces. McCreary refused and repeated his demand for absolute obedience to his orders. Union official O'Donnell asked, "Can we meet privately for a while?" The sheriff consented. After a few minutes in private session, a senior committee officer barked, "He ain't bluffing. Our asses are on the line. We got to do something." In response, the committee formally resigned to negate personal responsibility. Informally, they still controlled Homestead.

The sweltering July heat fueled labor to the boiling point. These workers had swallowed pain, abuse and low wages all their lives. Now, Frick and Potter had stolen their jobs. Signs around the fence threatened arrest, fine and imprisonment for anyone interfering with scabs. Desperate men pursued desperate measures.

That night, around 10:30 p.m., Superintendent Potter and Deputy Joseph Grey from the Allegheny County sheriff's office secreted to the Davis Island Dam in Bellevue under darkness to rendezvous with Pinkerton deputies—an assortment of imported ex-soldiers, skilled lawmen, rookies and thugs. Toward midnight, Captain Frederick Heinde loaded his armed Pinkerton troops aboard the barges *Monongahela* and *Iron Mountain*, which had been outfitted with bunks and a dining area. The tugboat *Little Bill* towed the barges toward Homestead.

The Pinkerton detectives rested through the night despite the annoying chug of *Little Bill*'s steam engine. As the flotilla approached Lock 1 on the Monongahela, a union scout perched atop the Smithfield Bridge telegraphed headquarters: "Watch the river. Steamer barges left here."[110] As the Pinkerton forces neared the plant wharf, sentries alerted to the intrusion fired warning shots. An unemployed laborer screamed, "Halt, damn you. Don't steal our jobs."

Frick had promised Carnegie to stay within the law, but he ordered his troops to answer violence in kind. The union launched an armed steamer, the *Edna*, to intercept the barges. Gunners unleashed a warning salvo of cannon fodder across the bow of the tugboat. A rifle shot from the shoreline shattered *Little Bill*'s pilothouse window. Unseen militants threw rocks toward the barges to prevent their docking. Shouts of "Don't land" echoed from the shore. Union leader Hugh O'Donnell pleaded, "In the name of God and humanity, don't attempt to land."[111]

The helmsman of the *Little Bill* ignored the cries from shore and docked at the wharf. Captain Fredrick Heinde, decked out in a navy dress jacket with shiny brass buttons, stood tall on the *Iron Mountain* and barked at the mob, "We were sent here to take possession of this property and to guard it for the company. We don't wish to shed blood, but we are determined to go there and shall do it."

The cry "Stay on the barge" reverberated from the shore. Neither side made a move until daybreak. A thick fog blanketed the riverbank, hampering a landing. Around 7:00 a.m., Captain Heinde warned the protesters of his intention to enter the plant. At 7:45 a.m., a Pinkerton threw down a gangplank, and Heinde crossed. Laborer Billy Foy hurled his body in the captain's path, receiving a whack from Heinde's billy club for his troubles. A protester in a red plaid shirt slapped Heinde in reprisal. A shot rang through the air. No one knew who fired first, but the powder keg of violence exploded.

A hail of bullets from the barges roared through the air. One shot sliced into defender Billy Foy. Another bullet from an armed striker dropped Captain Frederick Heinde. A projectile grazed Hugh O'Donnell's thumb, drawing blood. Thirty-four-year-old Joseph Sotak, who toted a loaf of bread home for his dinner, spotted wounded unionist Martin Murray: "You're wounded, man. Let me help you." As he dragged Murray to safety, a stray bullet shattered his jaw. Sotak spat out a mouthful of teeth and hefted the blood-soaked loaf above his head, mumbling through the agony of his death throes, "You cannot take this from our mouths."[112] Pinkerton J.W. Kline bled out and died. A bullet struck Homestead native George Rutler in the thigh. He would die in agony eleven days later.

Despite their losses, the unionists repelled the initial Pinkerton onslaught. A lull in the action allowed each side to drag away their dead and wounded. The tugboat *Little Bill* evacuated the more severely injured Pinkerton guards to the opposite bank for medical attention.

The fog lifted thirty minutes later, around eight o'clock in the morning, and the Pinkerton agents reconnoitered for a new battle. Twenty-eight-year-old Welsh immigrant John Morris, a blooming mill laborer, popped his head from a window to view the violence, only to be struck through the forehead by a stray bullet. He toppled into a ditch below, dead before he hit the ground.

Dozens of angry citizens hurled rocks at the barges while screaming epithets. A burning creosote-soaked railroad tie thrown by the mob floated toward the *Iron Mountain*. A burst of chain-metal fired mortar-style from the opposite bank raked across the *Monongahela* but fell harmlessly into the river. A second blast of scrap iron and nail shot sailed over the barge and decapitated Silas Wain, an innocent who watched the fray from a distance. Nineteen-year-old teamster Henry Striegel tripped and wounded himself. A Pinkerton sharpshooter's bullet through his neck finished the job. Unarmed Slovak Peter Fares fell from a stray shot. Unable to land against so strong an opposition, the Pinkerton troops retreated to the barges.

By four o'clock in the afternoon, armed labor sympathizers and gawkers from the neighboring Duquesne, South Side and J. Edgar Thomson plants had swelled the angry mob to nearly five thousand. With the *Little Bill* still on the opposite bank, the outmanned and outgunned Pinkerton army lacked an avenue of escape. Two untested recruits cowered beneath their seats while sporadic cannon and mortar fire from the shore rained terror around them. A well-aimed bullet struck thirty-year-old Thomas Connors in the right arm, severing an artery. His face turned ashen as he began to bleed out. A burning railroad car aimed at the barges by the unionists missed the mark but spread panic. Barrels of flaming oil released in the river licked at the docks, and an exploding stick of dynamite frightened even the bravest Pinkerton. The stink of death and defeat spread through the barges. The battle had lasted nearly twelve hours. The slow and brutal death of Connors snuffed out the last breath of the Pinkertons' will to win. A uniformed recruit raised a white flag of surrender, but a solitary shout punctuated the respite in the fighting: "Don't let one escape alive!"[113]

Sheriff McCreary urged the union leaders to halt the violence, but matters had gotten out of hand. He fired off a telegram to Pennsylvania governor Robert Pattison requesting immediate aid: "Nothing short of the militia will quell this disturbance."[114] William Weihe, president of the AAISW, stood up and pleaded with his men to cease firing, but cries of "Burn the boats" and blasts of dynamite drowned his voice. Mob rule had defeated reason. The blare of rockets and the hiss of Roman candles amplified the fear and confusion on the barges. A wounded Hugh O'Donnell walked through his pack of wolves preaching restraint. At five o'clock in the afternoon, the *Iron Mountain* hoisted another white flag. A contingent of Pinkerton men raised their hands in surrender.

Word of the onslaught spread across the floor at J. Edgar Thomson. When the day shift ended at 6:30 p.m., a parade of workers headed to Homestead, drawn like a moth to the flame. The final cracks from rifle shots, moans of the wounded, threats and curses from the mob assaulted their ears. The gruesome sight of the fallen, wisps of smoke from spent shot, the ferocity of the crowd and the defeated Pinkerton forces huddled against the sides of the barges attacked their eyes. Most hung back and watched the final act like spectators at a Roman bloodfest, both attracted and repulsed by the violent play before them. Others joined the tormentors.

More than one hundred locals burst across the gangplank unopposed and boarded the barges. The mob confiscated weapons and looted everything of value. In the excitement, laborer Thomas Weldin seized a Winchester rifle, which accidentally discharged as he slammed it to the floor. Within minutes, Weldin died from the self-inflicted wound. Newspaper reporters chased behind, discovering the dead body of Connors along with fourteen wounded Pinkerton Hessians huddled in a corner like a brood of motherless piglets.

Onlookers from J. Edgar Thomson witnessed their militant brothers prod and push the surrendering detectives from the barges with knives, sticks and guns to the accompanying shouts of "Off, you scum!" and "Murderers!" After the removal of the wounded, wild-eyed rioters set the barges afire. Cheers and yelps rose from the shore as flames consumed the boats.

On shore, the mob split in half and formed a gauntlet of terror. The unionists wanted revenge. They demanded blood. As the hordes pushed the first Pinkerton through the line, Homesteaders taunted and tripped the intruder, kicking and slapping him. Women and children joined the fray, tearing the bright blue uniform and brass buttons from his body. With each ensuing prisoner, the torture grew more vindictive. Rocks pelted the prisoners from afar. Clubs crashed across heads and shoulders, and punches struck noses and ears. The pathetic captives staggered through the line absorbing punishment. Blood oozed from fallen bodies. Cries of "mercy" went unheeded. An undersized woman punctured the eye of a fallen Pinkerton with the point of her umbrella. More than thirty later were hospitalized for broken bones and lacerations.

Armed laborers forced those at gunpoint who could walk or dragged the more severely wounded into the Homestead Opera House or the skating rink. There they suffered continued verbal and physical abuse until well after midnight. A teenager mocked the prisoners: "How do you like these apples, scum. Maybe a kick in the ass will teach you a little respect for the working man."

The viciousness of human ripping into human sickened the more sensitive onlookers who came to gawk but not partake in the spectacle. The bosses were heartless sons of bitches, but that didn't justify the day's violence. The union could not win such a war. Thank God this had not happened at his plant, J. Edgar Thomson, thought one Ukrainian worker. As the man left the scene of battle, he pondered the events of the night. There would be consequences. Labor had won the battle, but the war might end badly.

City officials begged the captors to halt the brutality. After compelling a promise to institute legal proceedings against the Pinkerton guards, Hugh O'Donnell and the union advisory committee allowed the Allegheny sheriff to marshal non-hospitalized prisoners to the Pennsylvania Railroad yard for a morning transfer to Philadelphia. The city's labor supporters celebrated with alternating catcalls against management and cheers for the union heroes. They had carried the day, or so they thought. Carnegie officer Francis Lovejoy dulled the union's optimism with a pronouncement signaling the AAISW's end: "The Homestead mill will never again recognize the Amalgamated Association nor any other labor union."[115]

Although only two detectives and six strikers died in the mêlée, fewer than in the Morehead Coal riot the prior year, most of the two hundred Pinkerton combatants suffered severe wounds. The savageness of the confrontation garnered national headlines as the press churned out coverage. As victors, the AAISW believed they held the upper hand. With nearly one hundred newspaper writers roaming the city, the union censored any information crossing the telegraph lines.

The union defenders had routed the Pinkerton trespassers, but their savaging of the prisoners turned public sentiment against them. Frick confidently felt victory. Carnegie cabled his support on July 6 from Scotland:

"Know you stand firm. Never employ one of these rioters." Privately, Carnegie blamed Frick's militancy for the public relations disaster. Labor and the press condemned Frick as an industrial Satan. The Reverend John McIlyar of the Fourth Methodist Church sermonized, "One man is responsible—Frick. This town is bathed in tears today, and it is all brought about by one man, who is less respected by the laboring people than any employer in the country."[116]

Thick-skinned Frick ignored his critics and adhered to the company dogma: "While nobody could regret the occurrences of the last few days more than myself, yet it is my duty, as the executive head of the Carnegie Company, to protect the interest of the association. The matter is out of our hands now. We look to the sheriff to protect our property. The men upon our property are not strikers, they are lawbreakers."[117]

After the battle, events moved swiftly in the company's favor. At Frick's request, the governor called out the militia to maintain peace. More than eight thousand uniformed Pennsylvania troops under Major General George Snowden mustered outside the Homestead plant on the twelfth in a show of force. The sheriff and his deputies rounded up outside agitators and shipped them out of town.

On the sixteenth, Frick ordered the posting of a notice:

> *Individual applications for employment at the Homestead Steelworks will be received by the general superintendent either in person or by letter until 6 p.m., Thursday, July 21, 1892. It is our desire to retain in our service all of our old employees whose past record is satisfactory and who did not take part in the attempts which have been made with our rights to manage our business. Such of our old employees who do not apply by the time above named will be considered as having no desire to reenter our employment.*[118]

Only 150 out of a workforce of nearly 4,000 applied.

General Snowden warned the citizens of Homestead that he refused to brook disruptions of any type. To demonstrate a seriousness of intent, the militia proceeded with a full-scale armed drill on the seventeenth. The following day, the Carnegie corporate attorney, Philander Knox, prepared a bill of murder and riot charges against the union leadership. Authorities arrested union leader O'Donnell for the fomenting of violence.

With the militia in control, the AAISW pushed for compromise. Whitelaw Reid, the liberal publisher of the *New York Tribune*, at O'Donnell's urging, wrote Carnegie seeking mercy, but mercy was not in Frick's vocabulary. In a letter to Carnegie, he gloated, "We had to teach our employees a lesson, and we have taught them one that they never will forget." Carnegie replied, "Am with you to the end, whether works run this year, next, or never."[119]

The country's business community supported Frick. Judge Mellon applauded his victory over the "labor parasites." John D. Rockefeller fired off a congratulatory telegram. B.F. Jones now possessed a platform to negotiate his own labor contracts from strength.

Fate would deal a jolt to Frick. Bedridden, his wife, Adelaide, neared her due date to deliver her fourth child, and her weakened condition concerned Frick. "I have some papers requiring my signature. I shall try to come home early. Go back to sleep." He kissed his wife lightly on the forehead. "Have a nice day," she whispered. It was Saturday morning, July 23, just a few weeks past the Homestead confrontation.

Alexander Berkman, a swarthy, black-eyed, twenty-five-year-old radical Lithuanian Jew, and his lover, anarchist Emma Goldman, raged at the deaths of the labor martyrs during the Homestead battle. He considered Frick a criminal. Since the courts refused to convict him of murder, Berkman appointed himself as judge and executioner. He considered the deployment of a bomb but reconsidered—too random, too unpredictable. He would pose as a New York employment agent named Samuel Bachman with an inventory of scab labor, gain access to the company offices and kill Frick. The irony of the plan amused Berkman—death by a delivery of scabs.

That Saturday morning, clad in an ill-fitting gray pinstriped suit, Berkman presented his card to Frick's office boy at the Chronicle-Telegraph building on Fifth Avenue: "I wish appointment for presenting a plentiful list of labor available for immediate employment." After checking with his boss, the boy announced, "Mr. Frick will see you at two."

A few minutes prior to the appointed time, Berkman returned. Without a word, he barged past the office boy while Frick sat at his desk in conversation with fellow executive John Leishman. Berkman lifted a .38 revolver from his pocket, aimed and shot Frick in the back. Blood oozed from the wound as Frick slumped and fell to the floor. Pint-sized Leishman leaped to his feet and seized Berkman's arm. Frick raised his head and shouted, "Murder! Help!"[120]

Berkman pushed Leishman aside for a second volley, but Leishman regained his balance and deflected the killer's wrist. The bullet sliced through the right side of Frick's neck, lodging under his left ear. As Berkman prepped a third shot, Leishman grappled with him. A wounded Frick struggled to

Attempted assassination of Henry Frick by Alexander Berkman.

his feet and seized the assassin as Berkman fired again. The bullet lodged harmlessly in the ceiling. The victim screamed for help, blood flowing from his wounds. A carpenter, alerted by the pandemonium, rushed into the room and struck Berkman in the head with a hammer. Heedless to the blow, Berkman pulled a sharpened file from his coat and stabbed Frick in the back. The blow bounced off a rib, but blood poured from the wound. A second thrust struck his hip. As Frick screamed, Leishman tripped Berkman and knocked him off his feet. Two clerks barged into the room as rescuers. The entire attack lasted only a minute but drew gawkers and good Samaritans from throughout the building.

Summoned to the scene from the street below, a deputy sheriff took control. Frick, still conscious, his face an ashen gray, blood staining his white shirt and his beard caked red, muttered, "Don't shoot. Leave him to the law."[121]

Leishman helped Frick rise to his feet on quivering legs and eased him to a chair. The deputy seized the blood-soaked Berkman and strong-armed him toward the door. In a flash, the wiry anarchist removed a mercury fulminate pellet from his pocket and stuffed it between his teeth. As he prepared to bite down, the officer forced the explosive from his mouth, saving disaster for all in the room. A search of the killer's pockets revealed several detonating cartridges and a stash of anarchist literature. With the arrival of reinforcements, authorities threw Berkman into a paddy wagon and to an eventual sentence of twenty-two years at Western Penitentiary.

"Well, I believe I feel like fainting," blurted out a hemorrhaging, pasty-faced Frick. Bright-red blood dripped from his neck, thoroughly soaking his starched white collar. Helping hands eased him from the chair into a chaise lounge. Employees summoned Dr. Lawrence Litchfield from his lunch at the neighboring Duquesne Club. The doctor arrived in minutes. Without instruments, Litchfield stabilized the patient and called surgeon J.J. Buchanan from nearby Mercy Hospital. Frick faced his wounds with bravery, joking, "Don't make it too bad, Doctor, for you know, I must be at the office on Monday."

"This will hurt. I am going to give you something to dull the pain," advised the surgeon. Frick refused chloroform, allowing his neck to be probed without benefit of anesthesia. The first bullet came out with minimal resistance, but the second involved persistent probing. Frick endured the torture without complaint, even assisting to pinpoint the location of the lead in his body. The stab wounds required significant cleaning, suturing and dressing as well, but the patient seemed stable.

Despite his condition, Frick attended to the day's paperwork, seemingly oblivious to pain. He dictated a telegram to his mother: "Was shot twice, but not dangerously."[122] A second note to Carnegie assured his safety: "There is no necessity for you to come home. I am still in shape to fight the battle out."[123] He signed the morning correspondence and prepared a formal statement to the press summarizing his hard line approach: "This incident will not change the attitude of the Carnegie Steel Company to the Amalgamated Association."

Frick asked to be taken home around four o'clock in the afternoon. "Now, don't alarm my wife about my wounds. She is with child." The medics carried him on a cot to an ambulance nearly two hours after the assassination attempt. As he passed Police Inspector Sylvus, who guarded the front of the building, he announced, "I'll be back Monday."[124]

For the first few days, Frick slipped in and out of consciousness from blood loss. The one-hundred-degree summer heat complicated recovery. Luckily, a rib deflected the stab wound to the back, and the bullets missed vital organs. As his body fought off infection, the delirious patient dreamed that dear departed Rosebud had saved his life through her goodness. A strong will to survive fortified Frick, and he gained strength daily. Neighbors Andrew Mellon, George Westinghouse and Henry Heinz visited his bedside, providing their good wishes for a speedy recovery.

Carnegie cabled, "Too glad of your escape to think of anything." Mentor Judge Thomas Mellon prescribed "quiet and mental repose."[125] Even the AAISW advisory committee condemned Berkman's "unlawful act and tender our sympathy."[126] As he healed, Frick read the daily newspapers and business reports to stay abreast of the union's activities. He never would forgive this assault on his life.

• • • • •

Word of the Frick assassination attempt spread through the neighboring J. Edgar Thomson plant on Monday morning like a raging wildfire. By lunch, worker chitchat seesawed between a twinge of sympathy and justifiable retribution against a villain who deserved punishment. Like a Greek chorus, the men on the floor debated the facts as they saw them, dealing in half truths and suppositions. The workers held little love for their unfeeling boss, but in the aftermath of the latest violence, the most humane subscribed to a belief that God never could support cold-blooded murder.

• • • • •

As Frick healed, the health of his newborn son faltered. With the boy's death imminent, the family christened him Henry Clay Frick Jr. On August 3, the infant died of internal bleeding. That very day, the court filed charges of murder against Frick for the Homestead riots. On the fourth, the family held a private funeral followed by an interment at the Homewood Cemetery. Tears lined Frick's cheeks as Henry Jr. was lowered into the family plot beside his sister. Although Frick healed physically, the combination of a son's death, his wife's despair and the liberal media's continued portrayal of him as a cad carried him to the edge of depression. Even his icy exterior could not freeze out the depth of the emptiness he felt. The day following the funeral, forty-three-year-old Henry Clay Frick sucked in a breath of summer air and walked out the door of Clayton to face a solitary streetcar ride to his downtown office for a return to business. Carnegie cabled him from Scotland: "Hearty congratulations from all here on your return to the post of duty. Everything is right when you and Mrs. Frick are right."[127] Even Carnegie's kind words rubbed like salt against his wounds.

The cowardly assassination attempt, Frick's personal bravery and sympathy for the loss of a son turned him into a heroic figure—at least with management. Frick initially rejected the offer for personal police protection against possible labor retribution, citing, "I am perfectly capable of taking care of myself." However, continued threats on his life forced the board to hire armed guards. Frick tried to forget his woes with a restful fishing trip while Adelaide donned black and retreated into ill health. She never would regain the sparkle of her youth.

Frick hardened his heart against the foreign trash who tried to murder him and who manned the mills. He would starve the union into total submission. On August 31, Frick visited Homestead without incident and announced, "The strike is a thing of the past." He recognized the union had lost. While the courts waded through a stack of claims and counterclaims, AAISW leader O'Donnell told the press, "The attempted assassination of Mr. Frick has created a bad impression all over the country. I would recommend an almost unconditional surrender." He added a corollary: "The bullet from Berkman's pistol went straight through the heart of the Homestead strike."[128]

The J. Edgar Thomson furnace crew mourned the passing of any hope for a labor-management balance of power. Poor pay, substandard working conditions and long hours would weigh heavily against labor following the loss of the Homestead war. Management hired fifty scabs, housing the so-called black sheep in a ghetto the union laborers nicknamed "Pottersville." During the late summer, typhoid fever and dysentery racked the city of Homestead, further sapping the community's declining spirit. The unseasonably hot September made the mill a hellish inferno. Injuries soared among the untrained scabs. Some of the replacements suffered an undiagnosed malady, undoubtedly induced by a mysterious yellow powder snuck into their food by union dissidents, possibly rat poison.

A delegation of ministers and civic leaders begged Frick to rehire the union workforce, but he refused to forgive or forget: "If the president of the United States, the justices of the Supreme Court together with Governor Pattison came before me and knelt down on their knees in this matter, my answer would still be no."[129] On September 17, Frick treated himself to the purchase of *Mine Mule*, an oil painting by James Boner. The animal toted the sign "Non-union helper," symbolizing both his personal victory and utter contempt for organized labor. Frick hung the oil above Clayton's back steps, the very route on which the medics had toted him to his bedroom the day of Berkman's failed assassination attempt.

Labor blamed Frick for the murders of their brothers and the death of the union, judging Superintendent Potter equally guilty. Potter's continued presence at Homestead would chafe labor like alcohol on an open sore. As a result, Frick and Carnegie shunted him from line authority to a staff position as chief mechanical engineer, a lateral move culminating in his resignation the next year. Thirty-one-year-old Charlie Schwab accepted Potter's superintendent's post at Homestead after receiving an increase in salary and the added incentive of a small ownership position.

On the evening of the twenty-fifth, three union representatives called on Schwab at his home seeking compromise. The new superintendent instantly recognized weakness. Matters had progressed too far to deviate from the company policy of total union annihilation. He vetoed the request to rehire union members on any terms.

Carnegie's steel competitors joined the celebration of labor's defeat. Henry Oliver, Ben Franklin Jones and the titans of the Duquesne Club toasted Frick's gutsy moves over cocktails and snacks. One gleeful executive snickered, "I guess old Frick kicked their hind ends. Now, isn't that just dandy." The Homestead annihilation provided the final proverbial nail in the AAISW coffin. The labor war had cost the Homestead plant profits of $300,000, but the workers, who could ill afford a single dollar's loss in pay, absorbed $1,200,000 in lost wages. Labor had won a battle but indeed lost the war.

Schwab induced the non-union mechanical department and a dozen skilled craftsmen from the rod department to return. A craftsman immediately picked a fight with an African American scab. The eruption burst into a full-scale mêlée, with both sides brandishing knives and guns. Management summoned the police to halt further bloodshed.

After several months of unemployment, on November 18, 1892, 500 former strikers snaked their way through the Homestead gate as non-affiliated workers while Schwab and Frick silently watched. The company had blacklisted 160 troublemakers, declaring them ineligible for rehire. A personnel agent checked every worker before granting permission to return, demanding an affidavit certifying that he had not taken part in violence. Two days later, 192 AAISW holdouts attended a final union meeting and voted 101 to 91 to return to work. "We surrender with both hands up," announced AAISW treasurer William Gaches.[130] Management allowed those not blacklisted to return as nonunion individuals.

Schwab now led a plant physically and spiritually in shambles. He advised the board, "The works were badly run down, and the men unsuited to their work, and they did not have competent foremen. The first four months of my time at Homestead were devoted entirely to reorganization."[131] Schwab lived in a company home and rarely left the grounds, working stretches up to seventy-two hours with only short naps to maintain his strength. He spoke with everyone, urging, cajoling and commiserating as only the best leaders can do.

Schwab warned Frick, "The converting mill is in terrible condition, and with the present machinery it is going to be a difficult matter to increase product."[132] The ordeal of reorganization had wearied the new superintendent. He needed downtime. Frick suggested, "Why don't you go to Scotland and present your proposed improvements to Andy in person?" Schwab welcomed the opportunity, and Carnegie greeted him like a long-lost son: "Charlie, you've stabilized Homestead like no one else could do." He readily authorized the needed expenditures.

New technology cut production costs by $500,000 annually at Homestead through the installation of labor-saving devices such as the automatic furnace loader. The annual expenditure for drinking water for the workforce ran $8,000 to $10,000 per year. Charlie located water from a "well we have at hand" and saved a bundle of cash. In short order, Schwab bragged to the board, "I beg to say that reductions we have made are the lowest I possibly can make."[133] Carnegie crowed to Frick about his supervisor: "Schwab is a genius in the management of men and machinery. I never saw a man who could grasp a new idea so quickly." The men on the floor at Homestead liked the new superintendent even though he adhered firmly to the company mantra: "I will not permit myself to be in a position of having labor dictate to management."[134]

The bosses assumed the animosity at Homestead had bottomed. Labor knew differently. They quietly stomached daily peril over long hours at bare-subsistence pay. These steel drudges held no loyalty for their stingy employers. They waited with revenge in their hearts—holding out for justice as compensation for the abuse the union had endured. Military armor plate contracts would provide the defeated unionists with just the vehicle.

Schwab's duties at Homestead included the supervision of navy armor-plate contracts. In September 1893, nearly a year after the strike, four disgruntled workers approached Secretary of the Navy Hilary Herbert, citing proof of fraud on military armament contracts. In exchange for a cash reward, they provided proof of falsified test results and documentation certifying to the unauthorized retreating of plates. Herbert impaneled a three-man review board to examine their evidence, which convicted the company in absentia without allowing an opportunity to refute the charges. Herbert unilaterally fined Homestead 15 percent of the value on all armor plate previously shipped.

Andrew Carnegie fumed at the high-handed vendetta based on evidence from a handful of disgruntled whistleblowers. He appealed to President Grover Cleveland, who reduced the fine to 10 percent, or $140,484.94. Failed vice-presidential candidate and liberal newspaper-publisher Whitelaw Reid called for a full-scale investigation of fraud. Blame centered on Superintendent Schwab, who countered that the navy's standards were unrealistic and its inspectors untrained. The newspapers shot bullets through Schwab's explanations and painted him as an unpatriotic thief with a "disregard of truth and honesty."[135] The Carnegie management team treated Schwab as a victim rather than a culprit. When Schwab tendered a resignation to save the company further embarrassment, Frick refused it, citing his profitability statistics as support for retention.

Despite his youth, the pressure of the job and adverse publicity had wrecked havoc on Schwab's body and soul. A few weeks' fishing in Canada provided a temporary tonic. He returned to work rejuvenated. Although he held down labor costs, he supplied a smile and a pat on the back with his kick in the rear. In response, the workforce rallied to his side. One foreman spoke his mind: "Charlie ain't such a bad guy. The men like his style." Carnegie and Frick likewise supported Schwab's tenure at nonunion Homestead, and the board of managers elected him to its membership.

CHAPTER 4

BAD TIMES

The steelworker awakened that Wednesday morning, August 21, 1895, with a start—something felt terribly wrong. Thunder shook the house. He looked out the window from his bed to a clear sky—no wind, no rain. His wife slept through the noise, oblivious to the interruption. She lay quietly, weak and pale. Flecks of gray streaked her dark hair. A faint line crinkled her brow, and her body had thickened ever so slightly. Yet she still possessed that angelic glow he remembered from the first time he saw her almost two decades earlier. The woman had tossed and turned through the night with a low-grade fever—probably the start of the flu. The husband felt her forehead, warm and clammy to the touch. He should let her sleep, but she lifted her body to face him and tried to rise.

"Stay in bed a while, Koxaha," the worker whispered to his wife, using the Ukrainian word for sweetheart. "Our son still in bed." He placed a hand on her shoulder. "Go back sleep."

"Okay," she managed weakly. A bead of sweat dripped from her brow. The steelworker swatted an annoying mosquito buzzing around her unprotected head.

His wife had packed ham-bologna from the meat shop, fresh brown bread and two apples in his lunch bucket last night. He would eat one of the apples on his walk to work.

"Almost 5:30, I got to leave. Feel better, and I see you around 6:30 tonight."

"I'll be fine by then. Have good day." The wife's head and stomach ached. As she scratched her hip, she noticed a tinge of faint red tattooed across her stomach. She coughed twice. Once her husband left, she pulled the blanket across her chest, shivered and closed her eyes.

The steelworker's apple tasted fresh and sweet as he marched to work. Random thoughts poured through his mind. His teenage son was getting so tall. He hoped the boy could avoid the furnaces. The men had pushed output to near maximum capacity. The foreman had congratulated the crew for a job well done, but the money was poor. As the worker neared the J. Edgar Thomson plant, a flurry of activity seized his eyes. Men and women scurried in all directions like tiny insects. Shouts filled the air, and the roof of the plant vomited thick blood-red smoke. The whistle had yet to signal the start of the morning shift. The newcomer nudged a coworker. "What go on?"

"Good God, it's a disaster. A hang fell in furnace H about an hour ago—huge explosion," blurted the fellow, gasping for breath.

So that was the jolt awakening the steelworker at five o'clock this morning. As he learned, a Hungarian laborer had dumped a load of stock into the furnace without releasing pressure—a rookie move. When a slag hang dropped and blocked the air escape pipe, gas exploded and burst through the top of the furnace with tenfold the intensity of an artillery strike.

"Many hurt?" asked the worker.

"Shit yes, but ain't got no names yet," exclaimed the coworker.

The man joined the crowd that raced onto the floor without punching in. The veins on his neck pulsated with each step. Sweat poured from his brow. A whining chorus of groans and shrieks greeted his entry. Clouds of smoke burned his chest. "God, help me," pleaded John Wagosky, a man the worker knew. Two volunteers had dragged him to safety. He was one of the lucky ones. He would live.

Word of the Armageddon had flamed across the community. Wives and family of the Slavic and Hungarian laborers joined the steelworker and forced their way into the mill, ignoring orders from the bosses to leave for their own safety. Few spoke English or understood. Those who knew but a handful of words cared only for the welfare of their loved ones.

Pandemonium reigned. The explosion had shattered the furnace top with volcanic force, blowing through the roof. Rescuers carted the last of the injured from the steel walkways. Charred skin bubbled up and peeled off a man's contorted face as two rescuers carried the semiconscious victim down the parapet. Even though the brave volunteers wore gloves, scalding metal railing seared their fingers. The catastrophe might have been worse had the explosion breached the side of the furnace, which would have poured waves of death horizontally across the plant. One poor soul had died during the first blast, tossed to the floor like a rag doll, burned and crushed beyond recognition. A handful of mutilated victims expired while being taken to the shed for medical treatment, some gone within minutes.

Whispers of the names of the dead and dying passed from mouth to mouth: forty-year-old Joseph Luckai, a father of four; Austrian Stephen Havrila, his head sliced open like a lemon peel; John Propokovitch; James Grucha; Joseph Cot; and John Mika. John Warha, John Skomda, Michael Jura, Andrew Droba and Michael Koperos would die within the week at Mercy Hospital. Howls of mourning women intermingled with the cries of the wounded. Several men lay comatose, insensible to the severity of their mangled limbs and disfiguring burns.

Steelworker.

The bosses shut down the plant and sent the men home. The husband found his wife still in bed where she had been hours before. She seemed too sick to take in all the details: "Husband, I sorry, but I must rest." Later, he heard the company gave seventy-five dollars to each family for funeral expenses. So that was the worth of a man—seventy-five dollars. The explosion could have killed him rather than the men of furnace H. He had passed fifty, ancient for a steelworker. His wife could have received seventy-five dollars instead of Joseph Lucka's widow. The steelworker touched the scar on his face, a grim reminder of his own close brush with death when a loose chain sliced through his jaw.

The steelworker had known Lucka, and he attended the funeral a few days later. His wife remained at home. "I not strong enough to leave house today," she begged off. Her fever had intensified, and the rash colored her cheeks. The husband summoned the doctor after his return from the funeral.

The woman lay motionless on the bed. She looked hurt and helpless. Redness blotted her upper chest and assaulted her face like a series of disconnected puzzle pieces. Her stomach rumbled and had grown distended. A tiny drop of blood oozed from her nose. She awoke at the doctor's touch as he felt her brow. The bedroom smelled of sickness. The doctor recognized the telltale signs at once—typhoid fever, the scourge of the nineteenth century. Some believed tainted drinking water bred the disease. Others pointed to rats, mosquitoes or rotten food.

The wife was very sick. She needed sleep, water and prayers, the doctor advised. She still was young, and most people her age overcame typhoid. The odds were good, but she showed no improvement the next week. Her nose bled again, and her stool took on a ghastly green color.

During the third week, the woman deteriorated rapidly. She bled from the rectum and refused food. She tossed and turned. Her eyes remained shut. She muttered incoherently. The distinctive rash faded, but her face took on a pale, waxy cast. She was dying.

A single tear dropped from the husband's cheek. He would not cry. Men did not cry, but the sadness crushed him. Life in Braddock was hard, but his wife possessed the gift to bring a smile to his heart. He loved the cute lilt of her Polish accent, the way she had difficulty with certain words. In bad times and good, she lifted his soul. At night, in bed, she was soft to the touch, sweet and loving. Her tiny breasts had given comfort and made him want her as much as when she was a young bride. She had an angelic goodness about her. She had worked part time at the grocery store to add a few dollars to the family coffers, and she rarely asked for possessions for herself—maybe a dress or material for curtains. Her love more than compensated for the howl of the furnace, the hiss of the fires, the threat of fall and injury.

The priest arrived in time to deliver the final rites. Neighbors brought food following the funeral, a blur of prayers and confusion. The husband picked up the Bible seeking solace. He read Proverbs 31, verse 10:

> *A woman of valor who can find her? Far beyond pearls is her value.*
> *Her husband's heart trusts in her and he shall lack no fortune.*
> *She repays his good, but never his harm, all the days of her life.*
> *She seeks out wool and linen, and her hands work willingly.*

The distraught worker repeated to himself that men did not cry. God must have a plan for his wife, but how could he take this angel from him? No other woman could take her place. All that remained were her memory and his son, who recently had turned thirteen.

The bosses understood why the laborer had taken the day off and had excused the absence. Many unfortunates died from typhoid that summer. On his return to work the day after the funeral, the foreman patted him on the back and expressed his sympathy: "Sorry about your loss." When a co-worker saw he had forgotten to pack a lunch, he handed him half a sandwich to tide him over.

For several days, the man moped around the house after work like a puppy missing its master, oblivious to his surroundings. His son begged him to snap out of his funk. "Mama would have wanted us to keep going. Papa, I miss her, too, but you have to eat." The man barked at his son but immediately regretted the outburst. His son was right. He worked at the grocery store and soon would enter his freshman year in high school. He was a good boy, and the father wanted him to complete his studies. No one in the family ever had finished high school. Even the great Andy Carnegie had never finished.

The worker understood he had to watch over his son. The life of a steelworker was brutal, but that is just how life was. "Toughen up, you have a son to raise," he told himself. He would miss his wife but refused to allow himself to surrender to despair.

CHAPTER 5

THE RIFT

Carnegie and Frick represented polar opposites—two egoists bound on a collision course. Carnegie spoke his thoughts aloud; Frick carefully thought out each word before speaking. Carnegie pursued public adulation; Frick sought anonymity. Few realized Frick had paid for the education of the children of two prominent Homestead strike leaders. Everyone knew about Carnegie's gifts of libraries and church organs. Carnegie chased the big idea, sometimes holding fast to a preconceived conclusion. Frick weighed each detail before establishing a firm position.

• • • • •

When Tom Carnegie's boyhood pal Henry Oliver proposed an investment in Lake Superior Mesabi iron ore, Carnegie rejected the idea out of hand due to bad past experiences with both ore and Oliver. Carnegie recalled former partner Andrew Kloman plunging into ore futures years earlier—a mistake that nearly torpedoed the company. He rationalized, "Oliver's ore bargain is just like him—nothing in it. If there is any department of business which offers no inducement, it is ore. It never has been profitable, and the Mesabi is not the last great deposit that Lake Superior is to reveal."[136] Then there was Oliver's can't-miss tip to buy a majority interest in the Pittsburgh and Western Railroad linking Allegheny City to the growing New Castle market. That investment error cost Carnegie thousands of dollars.

Henry Oliver was a promoter who concealed his Slabtown roots in the trappings of a fine gentleman. Frequently light on cash, he paraded around the city according to this corollary: act rich and people will think you rich. To customers and business acquaintances, appearances often masked reality. A silk top hat, Poole frock coat, starched collar and meticulously knotted necktie projected just the right panache. Oliver's Ridge Street home in Allegheny City sparkled with opulent touches. The *Pittsburgh Press* lauded the mansion as resplendent with "the best paneling, carving, wood inlay, and marble available." A properly attired maid served the evening meal to the Oliver family and guests in one of the finest mahogany dining rooms of its kind in the county decorated with "*fleur de lis*, geometric motifs, flowers, and leaves."[137]

From a humble start in life, Oliver had achieved a significant level of success. As a child, his mother fretted that he couldn't drive a nail straight, and he lacked the manual dexterity to help his father with the chores at the small saddle shop he ran. Henry's mother wondered if her son would ever earn a living, but a visiting phrenologist assured her that Henry was a bright lad and would find his place using his brain instead of his fingers. The phrenologist was right. This former Pittsburgh city councilman missed the Republican nomination to the United States Senate by an eyelash and co-owned several iron, wire and nail companies. He served as a past president and founding father of the Duquesne Club and counted neighbor iron and steel magnate B.F. Jones as one of his closest friends. In fact, Carnegie viewed Oliver as a younger and poorer version of himself.

In August 1892, following the Republican National Convention in Minnesota, where he had served as a delegate, Oliver took off on horseback with a few other curious Pennsylvania convention-goers to check out the excitement surrounding the Mesabi iron ore fields. The mines percolated with the same type of frenzy that had greeted entrepreneurs during the California gold rush forty-odd years earlier. Caught up by the vast potential he witnessed, Oliver kited a $5,000 check on a closed Pittsburgh bank as a good-faith down payment toward a promised $75,000 royalty for an ore procurement contract. His Irish-born assistant, George Tener, gasped in disbelief as Oliver handed over the check to the mine owner at a time when a financial squeeze strangled his iron companies in a vice-like lock: "Are you daft? You are stone broke!" Oliver slowly exhaled the smoke from his cigar, punctuating his answer with a coy Mona Lisa smile: "We shall see, Mr. Tener. We shall see." As soon as he returned to Pittsburgh, Oliver called on a friendly banker who covered the overdraft with a loan financed by a lien on one of his ore boats.

The cyclical ebbs and flows of manufacturing had squeezed Oliver. His iron business was mired in the throes of receivership. Minnesota's Mesabi iron ore offered a road to recovery, and the country's largest iron ore user, Carnegie Steel, represented his best opportunity for a partnership. He approached Henry Clay Frick to promote the mutual advantages of an investment in the Oliver Mining Company.

Carnegie considered Oliver a "plunger," a speculator who leaped into new investments willy-nilly. Frick, still weak from the prior month's assassination attempt, listened patiently to Oliver's presentation without betraying a hint of interest. After separating the facts from the sales malarkey, he felt the pitch made sense.

Frick eventually convinced Carnegie and the board in the wisdom of a 50/50 ownership in the Oliver Mining Company in exchange for a $500,000 loan, secured by a mortgage on the ore property. The deal would generate millions of dollars in ore savings through the next decade and make Oliver a very rich man. Steel historian Herbert Casson aptly dubbed Henry Oliver a knight errant with "the luck of a man who falls out of a three-story window and invents a flying machine."[138]

Once the investment began to prove itself, Carnegie jotted a note to Frick from Italy: "Oliver hasn't much of a bargain in his Mesabi, as I see it, but it is a good policy to take the half as a wedge that can be driven in somewhere to our advantage."[139] The partnership protected Carnegie against America's largest iron ore purveyor and a potential entrant into the steel manufacturing market, Cleveland multimillionaire John D. Rockefeller.

Frick believed what was good for his companies benefited the country. He detested second guessers in his affairs and shared his thoughts only with a handful of trusted friends. Superintendent Charles Schwab considered him a total enigma: "No man on earth could get close to him or fathom him."[140]

In April 1894, labor unrest roiled the Connellsville coal fields. Roving gangs of union malcontents bullied management, who responded with violence. Three workers died in one such confrontation. The strikers answered with the murder of H.C. Frick engineer J.H. Paddock. Frick knew but one solution: attack hard and without mercy. Labor quickly caved when faced with an army of mercenaries. Carnegie questioned Frick's brutal use of force but held his tongue.

Each week, some issue drove a wedge deeper between the executives. When Frick entered into secret negotiations with Russia for a large steel plate contract, Carnegie blabbed to the press. Using the information revealed, rival Bethlehem Steel undercut the Carnegie bid and stole the contract. Frick complained to the board, "It is unfortunate that our leading stockholder is a little injudicious at times"—a huge understatement. Carnegie's loose words galled the uncommunicative Frick.

When Carnegie discussed a potential merger with an H.C. Frick Coke competitor without informing him, Clay grew livid. Nothing galled him more than interference in coal company affairs. Furious at such a betrayal, Frick resigned his position from the steel company on December 17, 1894. As a gentleman, he tendered two weeks' formal notice. Carnegie apologized and tried to soothe his partner's bruised ego, but to no avail. Frick held firm in his resignation: "The decision to retire from the firm was made after the most serious consideration, and I see no reason to recall it, and while the time is short, it is my duty and will be my pleasure to advise my successor in any way that I can."[141]

Partner Henry Phipps counseled Frick to reconsider and urged Carnegie to make amends: "We need Clay Frick. He is a profit maker." Carnegie apologized. The possibility for a reconciliation appeared in the offing until Frick accidentally encountered a letter from Carnegie referring to him as a "disordered man."

The blunder set off a new tirade by Frick, who wrote a January 1 threat-filled response:

> *If ever a man penned a lot of nonsense than you did when you wrote that letter to Mr. Phipps—copy which you sent me (by accident)—I should as a curiosity like to see it. It is high time you should stop this nonsensical talk about me being unwell, overstrained, etc. and treat this matter between us in a rational businesslike way. If you don't, I will take such measures as will convince you that I am full competent to take care of myself in every way. I desired to quietly withdraw, doing as little harm as possible to the interests of others, because I had become tired of your business methods, your absurd newspaper interviews, personal remarks, and unwarranted interference in matters you knew nothing about, I warn you carry this no further with me but come forward like a man and purchase my interest and let us part before it becomes impossible to continue to be friends.*[142]

Henry Phipps continued to champion Frick's return as CEO. He cited Clay's exceptional performance. Most importantly, he had delivered a steady dose of dividends. Phipps liked dividends, lots of dividends. Other partners seconded a reinstatement as well. However, Carnegie already had promoted John Leishman to Frick's post shortly after the resignation.

With Phipps's constant nudging, Carnegie apologized again and concluded a truce. Since Leishman now held the presidency, the company carved out a new executive position: chairman of the board of managers. After the executives shook hands, Frick assumed his senior partner would halt his future interference.

During 1895, the Connellsville Coal Syndicate instituted an across-the-board pay-rate reduction. Frick had warned the steel company in advance to stock additional coke inventories for the strike he knew would follow, advice overlooked by the purchasing agents. With profits reaching record levels, Carnegie refused to risk a raw material disruption. He ordered Frick to settle, again undermining his CEO's authority. Frick gritted his teeth in silence and chalked up another insult from his senior partner. Despite the friction between the partners, the steel company continued to rack up record profits.

By employing a combination of collusion, cutthroat competition, technological superiority and political and social connections, the Carnegie plants had seized the lion's share of the country's steel business. Low pricing generally won the contract. On other occasions, the strategy of dividing orders with Duquesne Club cronies boxed out weaker rivals and guaranteed huge earnings for the members of the trust. Labor gathered like hungry dogs while their masters gorged themselves at the banquet table, generating a massive backlog of resentment across the plant floor.

The Connellsville mine owners pushed out production with even less concern for their manpower than the steel plant execs. The addition of new blasting explosives in the mid-1880s further amplified risk. Airborne coal dust and mine gas presented the potential to spark and explode from a shovel striking rock or a malfunctioning oil lamp. Workers lucky enough to survive accidents, typhoid fever, pneumonia and dysentery rarely lived beyond their sixtieth birthdays due to black lung disease or physical collapse from the exhaustive work. Frick treated his foreign-born miners as replaceable commodities just as Carnegie did in his plants.

In contrast to the armor-plated exterior he displayed toward labor, Frick showered his family with warmth and decency, although son Childs, who would choose a career as a paleontologist rather than as a businessman, found his father difficult on more than one occasion. The loss of two children—especially Martha, whose likeness he engraved on his personal checks—had sucked the joy from his life. Frick promised to find more time for his son Childs and daughter Helen. He wished he only possessed the magic to make his wife smile again.

• • • • •

The workforce faced most days without heart. The men struggled at home with bills to pay and chores to do. Those with wives and children hoped education might protect their sons from the curse of the mills and mines. Although they had lost hope of a better life for themselves, they still held firm to the American dream for their children.

Laborers frequently obtained part-time work for their teenage sons, adding a few dollars to the family coffers, but they also pushed education. If someone higher up took a liking to the teenager, he might move up the chain, especially if he spoke without an accent and stayed in school past the eighth grade.

• • • • •

In March 1897, John Leishman resigned following a conflict-of-interest blowup pertaining to his partial ownership of a key vendor, a practice employed by Carnegie but one he denied his junior partners. Frick lobbied President McKinley for Leishman's face-saving appointment as ambassador to Switzerland, a repayment to the man who had saved his life.

The following month, Carnegie promoted eighteen-year veteran, thirty-five-year-old Charles Schwab to the company presidency at a $50,000 annual salary with an increased stock ownership kicker. William Corey replaced Schwab as Homestead superintendent. Alexander Peacock served as vice-president, Lawrence Phipps (Henry Phipps's nephew) as second vice-president and Francis Lovejoy as secretary. Frick continued as chairman of the board with broad undefined corporate powers. Henry Phipps and Dod Lauder served as lead board members.

Schwab seized the operational reins at a gallop, carrying out an aggressive agenda. He warned the board of vast changes on the horizon. "Market demand has shifted from rails to skyscrapers and bridges. We have the greatest difficulty in getting our customers to take Bessemer structural steel."[143] He pointed to the growing strength of competitor Illinois Steel. He proposed the construction of sixteen modern furnaces and a new blooming mill at Homestead, along with a major remodeling of the Duquesne plant. Conservative Henry Phipps gasped at the enormity of the program, but

Schwab possessed little patience for Phipps, a man whose time had passed: "He had no vigor or initiative and was of the bookkeeper type."[144]

Despite the objections from older board members like Lauder and Phipps, who pursued higher dividends rather than growth, Carnegie approved the suggested overhaul. When Homestead's cost per ton dropped 34 percent, Carnegie congratulated his president: "You are a hustler." Even Phipps applauded Schwab.

Times were good. Carnegie and Frick had outwardly reconciled. When Carnegie donated funds for a new Pittsburgh art museum, Frick accepted the treasurer's position and contributed an oil painting. The partners served on the committee to replace Frederic Archer as conductor of the Pittsburgh Symphony and together pushed the blooming political career of corporate counsel Philander Knox. Frick also worked well with Schwab. The duo purchased the Carrie Furnace and constructed additional rail spurs, but even as the company achieved new highs, Carnegie's public pronouncements continued as an embarrassment to Frick. The Carnegie plants had earned hundreds of thousands of dollars by producing armor plate for the navy. Yet Andy Carnegie publicly ranted against war and opposed the U.S. annexation of the Philippines following the end of the Spanish-American hostilities. Frick strained to conceal his impatience at his senior partner's hypocrisy.

Schwab continued to perform admirably, and Carnegie gloated about his protégé to Frick: "His administration is a success, decidedly so, and I hope you feel as I that we have the right man in the right place, also that he relieves you from much anxiety, as he certainly does me. I've never felt so happy and contented in regard to our business as at the present time."[145]

Carnegie had reached sixty-five—an age by which most nineteenth-century men had either retired or died. One week he spoke of selling. The next he opted to expand. Frick, Phipps, Schwab and Judge Elbert Gary of Federal Steel openly had discussed the possibility of a buy-out, but talks collapsed after Frick refused to remain with the organization following any consolidation.

In April 1898, John Gates of American Steel and Wire broached a three-way merger with Illinois and Carnegie Steel, but Frick and Phipps thought the acquisition price inadequate. With Carnegie Steel removed from possible consideration, financier J.P. Morgan manufactured the $200 million Federal Steel Company, incorporating Illinois Steel, Lorain Steel and several coal properties. The new corporation rivaled Carnegie in breadth and depth. Andrew Carnegie recognized that the time had come to diversify or sell.

Phipps had dreamed of cashing in to enjoy his decades of labor. On November 23, 1898, Phipps and Carnegie found themselves in Arlington, Virginia. Carnegie had come to Washington to lobby against the United States' annexation of the Philippines. As the old friends strolled along a tree-lined boulevard, Carnegie put his arm around Phipps's shoulder and confided his own weariness: "Harry, we must be getting old. Do you think we should sell?"

"Andy, if we could get the right price, the time would be right."

Carnegie hinted $250 million might be a nice round figure, and Phipps concurred. Within the week, Phipps had shared the conversation with Dod Lauder and Louise Carnegie, who both agreed the time for a sale might have come.

Frick likewise endorsed the idea after a discussion with Phipps. He wanted loose from Carnegie's interference. At the February 20, 1899 board meeting, he sniped, "The senior has the last guess and the power to decide policy, which he does, frequently regardless of the board, but for sometime he has stumbled badly."

In March 1899, Frick and Phipps parlayed with corporate raider Judge William Moore of Chicago, receiving an offer to purchase the Carnegie Company for $250 million. Andy Carnegie would net $157,950,000. Frick and Phipps would split a bonus of $5 million in addition to their ownership shares. Moore demanded absolute secrecy regarding his involvement in the merger. Phipps liked the deal and pushed the proposal to Carnegie, who required either 58.5 percent of his share in cash or $1,170,000 for a ninety-day option. Since Moore could only raise $1 million on short notice, Frick and Phipps put up the $170,000 shortfall.

Problems arose quickly. A financial downturn forced the lead bank to renege on its loan commitment for the project. With the option nearing the termination date, Phipps and Frick steamed to Scotland and Carnegie's Skibo estate in search of more time to raise cash. Carnegie had assumed Andrew Mellon or John D. Rockefeller backed the offer. He detested speculators, whom he labeled in his *Gospel of Wealth* essays as "parasites fastened upon the labor of businessmen." When he discovered speculator Moore stood behind the offer, he refused to grant the syndicate a minute of additional time. Carnegie carped to the board, "Mr. Frick's partnership with Moore by which he was to make millions was a betrayal of trust."[146]

Carnegie pocketed the option money, including the $170,000 his junior partners had put up for the deal, saying that he called this windfall "a nice little present from Mr. Frick."[147] Carnegie had skunked Phipps and Frick.

The chastened partners slinked out of Skibo with their heads hanging. To Carnegie, taking Frick and Phipps's $170,000 had been a big joke. Phipps loved the dollar and saw no humor. Frick viewed it as a personal assault, and his grudge against Carnegie grew rancorous.

Frick attempted to form a new syndicate financed by Pittsburgh banker Andrew Mellon. The deal came close but collapsed. Mellon felt the wily Scot had toyed with him, writing years later, "Carnegie double-crossed me."

While Phipps and Frick had powwowed with Carnegie in Scotland, the AAISW attempted to make a comeback. The union took another shot at a Homestead union reorganization in May 1899, but the effort fizzled. Superintendent Corey fired four key labor agitators on the spot. Schwab advised the board, "This will nip the move in the bud. We will have no labor organization in our works."[148] Although he might be a good egg on an individual basis, Schwab proved a tough taskmaster and no friend to unionism. He implemented the company's firm anti-labor policies with relish, gaining his bosses' respect with each success, even as the friction between Frick and Carnegie simmered with hostility.

Frick's anger over his lost half of the $170,000 burned like pumice against a puss-filled sore. He seethed as the secretary read a letter at the September 11 Carnegie Brothers board meeting explaining the failed merger that ended with: "We do not take up with speculators."[149] The chairman vented by spewing invective regarding his absent partner's ill-conceived long-term rail contract rates, which had created a short-term price disadvantage for the company against its competition. Later in the month, Frick cooled his anger long enough to ratify a three-year verbal coke contract in New York with Carnegie at $1.35 per ton for the steel company, $0.15 below the prevailing market price.

Carnegie bragged to his cousin, "Dod, I think I just made a coup with the coke price I negotiated with Frick."

Board member Dod Lauder, who always called his cousin by his boyhood nickname, countered, "Aye, Naig, but what if prices decline?"

Carnegie rethought the deal and asked President Charlie Schwab to obtain a protective clause in lieu of a price drop. Frick refused to negotiate, assuming the Schwab counterproposal negated the recent oral agreement. As market pricing edged toward $2.00 per ton, Frick inched up the coke billing price to $1.75 per ton, still below the market rate. Schwab paid the coke bills as received but instructed the treasurer to mark Carnegie's records for the amount above $1.35 as "payment on advance accounts only." When the issue came up for discussion at the H.C. Frick October 25, 1899 board meeting, Frick commented, "Mr. Carnegie and I had a considerable talk about what the price of coke should be for, as he called it, a permanency." However, Frick believed the price issue had yet to be settled.[150]

When Carnegie reviewed the coke board's notes, he sniped to Frick, "Excuse me, I have no time to waste upon the pres. of the H.C. Frick Coke Company, who begins saying he didn't know the bargain. My friend, you are so touchy upon F. Co. We all have our crazy bones. Your partners will not speak to you freely about coke."[151]

Frick denied the existence of a contract: "I did personally agree to accept a low price for coke; but on my return from that interview in New York President Schwab came to me and said Mr. Lauder said the arrangement should provide that in case we sold coke below the price that Mr. Carnegie and I discussed the steel company was to have the benefit of the lower price. I then said to Mr. Schwab let the matter rest until Mr. Carnegie came out."[152]

Frick considered the brouhaha as a direct shot at his independence and integrity. Carnegie countered, "It's only business with nothing personal in it."[153] The Scot negotiated dollars and cents. With Frick, the matter had become personal. Carnegie's incessant meddling curdled his stomach. His media posturing caused a public relations nightmare. Nor could Frick forget the $85,000 Carnegie had taken from him as a result of the failed Judge Moore option.

No discussion pertaining to coal pricing came up at the November 6 board meeting with Carnegie attending. However, Frick offered the steel company over four hundred acres of prime industrial property he owned along the Monongahela River at Peter's Creek, just six miles from Homestead. He priced the land at $3,500 per acre or $1,500,000—$1 million more than he and silent partner Andrew Mellon had paid but less than the appraised value of $4,000 per acre. Carnegie sat silently and scowled during the proposal. When the meeting adjourned, he groused in the hallway to several board members concerning the impropriety of selling to one's partners at a profit.

When Frick heard scuttlebutt about Carnegie's accusation, he boiled. Inhaling deeply and slowly exhaling, he allowed his temper to cool. He would choose his time and place for a rebuttal, holding his tongue until the November 20 board meeting. With Carnegie absent, Frick rose to his feet and vented, "Why was he not man enough to say to my face what he has said behind my back?" For once he opted to speak before thinking. His voice gathered momentum and his face reddened: "Harmony is so essential for the success of any organization that I have stood a great many insults from Mr. Carnegie in the past, but I will submit to no further insults in the future." Board member and Carnegie cousin George "Dod" Lauder gasped but withheld comment as Frick unleashed a litany of complaints. At the completion of the diatribe, Frick stormed from the room.

Lauder immediately penned a letter to his cousin diagnosing the severity of the outburst: "To my mind the chairman seems to have deliberately burned his boats, and the issue is now Carnegie or Frick."[154]

Carnegie had had enough of Frick. He crafted a dissembling note of appeasement as a ploy:

> *My Dear Mr. Frick: Never did I say a word other than in praise of your conduct anent the purchase up the river. I said you got the property without effort. That it is doubtful whether a chairman could buy such land as his company might need, I could never say. This is just the result sure to come from tale-bearers—distortion—words innocent if one hears and knows the conditions and spirit in which they are said. I don't remember mentioning the coke difference except to those who said to me you had told them you would fix it with me. I am not guilty and can satisfy you of this—also the folly of believing tale-bearers—a mean lot.*[155]

Carnegie revealed his true intention to Dod: "You voice my views exactly. Frick goes out of the chairmanship of the board next election or before."[156]

Henry Phipps preached reconciliation, but events had progressed too far. Carnegie ordered recently promoted president Schwab to obtain Chairman Frick's resignation at once. Schwab found himself in rough water, caught between the opposing egos of his two bosses. Schwab railed to New York to reconfirm Carnegie's decision, since the two partners had shaken hands and settled their arguments many times in the past. Carnegie insisted that there would be no reprieve this time.

Schwab called on Frick at Clayton on December 4. He explained the "awkward situation" of having to choose sides and handed Frick a note opening: "My long association with you and your kindly and generous treatment of me makes it very hard to act as I shall be obliged to do." He explained that if the board had to choose sides, the company and the junior partners would suffer. Frick listened with restrained courtesy to the plea for his resignation. The chairman's detachment made his reaction impossible to decipher. The next day, the board received a brief note: "Gentlemen: I beg to present my resignation as a member of your board. Yours Very Truly, H.C. Frick."[157]

Frick saved the embarrassment of a firing, but Carnegie had loaded another broadside. Frick still owned 23 percent of the coke company, where he continued as chairman. He also held 6 percent of the steel company. Carnegie intended to ambush and punish Frick, writing cousin Dod Lauder: "He can't repudiate contracts for a company which myself and my friends control. We are not that kind of cat." Carnegie summoned the steel company board on January 7 and demanded the full deployment of the Iron Clad, an agreement that forced an inactive stockholder to sell his shares to the company at book rather than market value following an affirmative vote of 75 percent of the outstanding ownership.

At the H.C. Frick Coke Company, Carnegie fired three pro-Frick members from its five-person board and increased its size to seven. He replaced the old members with five of his cronies: Daniel Clemson, James Gayley, A. Morehead and cousins Dod Lauder and Tom Morrison. Only Thomas Lynch remained from the original board to side with Frick. On January 8, 1900, the new board voted to honor the disputed selling price of $1.35, despite the fact that the market had risen to nearly $3.50. In a further punitive adjustment, the board declared the contract retroactive, costing H.C. Frick $596,000 as a refund to Carnegie Steel—another slap at Frick's pocketbook. Frick listened in stunned silence as each motion passed. At the conclusion, Frick barked, "You will find that there are two sides to this matter." His face darkened, and he stomped out the door, hustling to meet with his attorney.

Carnegie called on Frick at his office the following day, as his former chairman cleaned out papers. Employing his utmost charm, Carnegie urged calmness to avoid legal messiness. Frick agreed, asking his prior partner merely to purchase his shares at fair market value.

"I can nae do that," answered the Scot, holding firm to book, a figure millions of dollars lower than market value. Carnegie pointed out that Frick didn't complain when John Leishman received book upon his departure from the company. Frick suggested they hire independent arbitrators to decide what is fair. Carnegie refused. Frick countered with an offer to buy Carnegie's shares at book. The Scotsman refused.

With all hope for compromise out the window, Frick turned ballistic. He leaped from his chair, his face contorted. "You're a God-damned thief without an honest bone in your body." With fists clenched, he bellied against his older partner and glared. Pointing his finger at Carnegie's nose, he threatened, "We will have a judge and jury of Allegheny County decide what you are to pay me."[158] No man ever had confronted Carnegie so aggressively.

The sixty-five-year-old fled through the door of Frick's ninth-floor office fearful of bodily injury.

Carnegie scurried down the hallway with Frick chasing behind him and screaming threats. The executive suite bustled with excitement. Philander Knox, soon to be the U.S. attorney general, peeked his head into the hallway and stared in amazement. Charlie Schwab eavesdropped from his office with dismay. As soon as Carnegie reached the safety of his own office, he summoned corporate secretary Francis Lovejoy and ordered, "Call an emergency meeting of the board at once." He had become more determined than ever to seize Frick's stock at book and make him pay for his insulting behavior. This outburst would be the last time Frick and Carnegie would ever speak to each other. The friendship of their wives, who once had been quite fond of each other, ended as well.

Later in the day, Frick's anger cooled even as his resolve stiffened. He confided to ousted coke director John Walker, "I lost my temper this morning." Walker answered with a grin, "Oh well, I knew you had one."[159]

At the January 11 Carnegie emergency board meeting, thirty-two members signed paperwork demanding Frick's shares at book. Square-shooter Francis Lovejoy refused, resigning his secretary's position rather than take advantage of Frick. A.R. Whitney and Henry Phipps likewise declined to sign.

Carnegie visited board member Henry Curry at his home on his deathbed in search of his support. Curry declined. "Why?" asked the surprised Scot. "Mr. Frick is my friend." "But am I not also your friend?" asked Carnegie. Curry replied, "Aye, but Mr. Frick never has humiliated me," referring to the demeaning manner in which Carnegie had criticized him on past occasions.

When Frick learned the result of the vote, he addressed a threatening letter to the board:

> *On Friday evening, January 12, 1900, for the first time, I learned that the board of managers secretly and without notice to me at a meeting on Monday, January 8th, 1900, passed a resolution offered by Andrew Carnegie. Mr. Carnegie thinks he can unfairly take from me my interest in the Carnegie Steel Company, Limited. Such proceedings are illegal and fraudulent, and I now give you formal notice that I will hold all persons pretending to act thereunder liable.*[160]

Just two days before the company threatened seizure of his shares, Frick reiterated a proposal for arbitration. B.F. Jones and George Westinghouse offered their services as mediators, but Carnegie adamantly refused.

Following the receipt of a February 1 letter from President Schwab demanding his shares under the Iron Clad, Frick instituted a suit in the Allegheny County courts, stipulating, "Carnegie without reason and activated by malevolent motive demanded his resignation. Your orator shows to Your Honor that this attempt of Carnegie to expel him from the firm and seize his interest at but a fraction of its value is not made by him in good faith."

Carnegie had underestimated Frick. If Carnegie wanted war, by God, Frick would oblige him. Henry Phipps, who fretted over the value of his own stock, allied himself with Frick. Phipps considered his boyhood friend and benefactor wrong on this issue: "It was understood that the 'Iron Clad' should only apply to debtor partners or employees. The agreement never was meant to be an instrument of oppression and robbery."[161] As the creator of the original Iron Clad, Phipps volunteered to dissect the agreement and uncover its weaknesses. Frick charged forward, unleashing a full salvo of cannonballs, while Phipps sniped with musket fire. Frick intended to attack Carnegie's character, painting him as a villain for all the world to see—a greedy predator who sold rails costing $12.00 per ton for $23.50 while paying his unskilled workers a mere $400.00 per year.

Frick hired renowned Philadelphia litigator John Johnson to represent him in court. Carnegie Steel counsels, Philander Knox and Judge James Reed, declined to represent the company, citing a conflict of interest. Carnegie settled on legal representation from the team of George Bispham, Richard Dale, Clarence Burleigh and Gibson Packer. The *Bulletin* wrote, "The event of the week just ending has been the opening of the battle of the giants—recourse to law by Frick to obtain what he regards as his rights as a stockholder in the Carnegie Steel Company. The history of Pittsburgh has failed to supply a parallel to the magnitude of this litigation wherein millions are at stake."[162]

Frick allies John Walker and Frederick Lynch simultaneously launched a separate suit on behalf of the coal company claiming Carnegie personally designed to "cheat and defraud, not honestly or in good faith for the coke company, but dishonestly and in bad faith for the benefit of said Carnegie."[163]

With the tenacity of a cornered raccoon, attorney Johnson attacked the Iron Clad Contract, producing a record of Frick-to-Carnegie conversations dating back to 1898. He argued:

> *The fact is that the present Iron-Clad Agreement which I believe is signed by all juniors (except those lately admitted) is not binding on any one of them, nor on anybody, and never has been, Mr. Vandevort never having*

> *signed it, and while we have acted under to date, purchasing interests of deceased partners under it, if there had been any objection raised on the part of their estates, it could have not been enforced.*[164]

Johnson pointed out that Henry Phipps had refused to sign the 1892 Iron Clad, further invalidating the document. To step up pressure, the Frick team threatened the release of Carnegie's private correspondence, as well as the company's profits, to the media—including the fact that Carnegie earned more in a single month than the president of the United States received in a year. Carnegie's advisers thought the public relations fallout could ruin his reputation and would incite a volley of anti-monopolistic legislation. Attorney Philander Knox "very strongly advised settlement with F. on any reasonable basis."[165] Board member W.H. Singer warned that the Iron Clad might fail in court. Interested industrialists and board members urged compromise. Recognizing the weakness in his position, Carnegie hauled up the white flag and parlayed at his New York home with Henry Phipps, Charlie Schwab and Francis Lovejoy to hammer out the terms of a truce. Carnegie agreed to accept a market-price valuation, providing that Frick would affirm his resignation and that all final dealings would flow through a trustee to eliminate the need for face-to-face negotiations.

The Carnegie board railed to Atlantic City, many accompanied by their wives to mask the intent of the meeting from prying eyes. On Thursday, March 22, 1900, Frick agreed to drop all legal action in exchange for $31,284,000 in securities and bonds rather than the $4,900,000 originally offered. When Carnegie Steel "readjusted" coke pricing, Walker and Lynch dropped their suit as well. In a final move, H.C. Frick Coke merged into Carnegie Steel to form a single conglomerate entity. Henry Phipps congratulated himself for his role since the projected value of his holdings reached an even higher figure than that of Frick: $34,800,000.

The valuation war had taken a toll on both former partners. Carnegie covered up battle fatigue with false bravado. While Frick left the field physically fit, he suffered emotional damage. His bitterness demanded further revenge. He resigned his position as treasurer of the Carnegie Museum and the Carnegie Institute. He next partnered with William Donner and Andrew Mellon to create Union Steel in Donora, Pennsylvania, a direct competitor to Carnegie Steel.

Carnegie received Frick's shares through an independent agent as agreed, avoiding contact between the former litigants. A few months later, Frick hurled a barb at his former partner by cable to Scotland: "You are being outgeneraled all along the line, and your management of the company has already become the subject of jest."[166]

On March 22, 1900, the board named Charles Schwab president of Carnegie Brothers, the successor corporation to the consolidated coal and steel companies, possibly the most important executive position in the country. Older board members like Dod Lauder and Henry Phipps continued to preach the wisdom of slow and steady progress—minimal investment coupled with maximum dividends. Phipps cautioned, "Let us develop what we now have—no groping in the dark. Why look beyond our own business?" The younger, more aggressive Schwab opposed the status quo. He warned the more timid board members, "If we want to drop back into an old-fashioned way of doing business I want to be counted out of it."[167]

Charlie Schwab oversaw all operations with Carnegie's full confidence. As chief negotiator, salesman and production supervisor, he counterbalanced Andy's speak-first, think-later mentality. He possessed the tact to cool down the pot whenever Andy's posturing caused a boil. He had a talent for wringing every ounce of profit from the production line—earning as a reward the huge annual salary of $100,000 plus 6 percent of the company stock. Charlie's broad smile and cheerful disposition tickled Carnegie—a pleasant change from Frick's dour disposition. "I hope you will remain as president until you are an old man." The boys in the plant liked "Smilin' Charlie" a lot more than old ice-water-in-his-veins Frick. "I'd follow him anywhere. He knows his steel," voiced one foreman. Schwab indeed had become the man at Carnegie Steel.

CHAPTER 6

U.S. STEEL

It was Sunday, December 30, 1899. The nineteenth century clung tenuously to its final hours. After Mass, a veteran steelworker sat in an easy chair by the fire with a book on his lap, awed by the swift passage of time. Although most of his cohorts at the plant could neither read nor write—knowing only how to craft the letter X—he had learned to read from the tattered *McGuffey Readers* he discovered in the attic of his landlady when he first arrived to the new country.

The heat from the fireplace warmed his chilled bones as vision after vision spun through his mind fanned by the reds, blues and oranges of the flames lapping at the logs. The years had flown past in the whirl of a tornado, punctuated by thunder and lightning—so many changes, some good, many not. He had seen and felt much over his lifetime. It seemed just yesterday he had been a child on a farm in the Ukraine, working the potato fields with his *batko* (father). Now, he labored at J. Edgar Thomson in Pittsburgh six days a week, besieged by the aches and pains from advancing age and plant injuries.

The world had morphed like Jules Verne's *Twenty Thousand Leagues Under the Sea*, the book sitting on his lap. The streets of Pittsburgh pulsed with George Westinghouse's electricity, Andrew Mellon's trolleys and Alexander Graham Bell's telephones. Dirt roads had become brick. The city of wood had changed to a city of iron and now a city of steel thanks to inventor Henry Bessemer. The science fiction of just a few decades ago had become today's reality. Although he never had used a telephone, the daily newspaper promised that nearly every home in Pittsburgh someday would own one—oh yes, and horseless carriages run by gasoline. Just last summer, Howard Heinz, the president of Heinz Foods, had motored a fancy French-designed Panhard automobile through the city streets. The steelworker had seen Mr. Heinz drive past him on a fall afternoon as he and his son took a Sunday stroll. They stared at the horseless carriage in amazement. So much had changed since he had left the farm as a teenager. Yes, he could be thankful for many things.

The steelworker's son would graduate from high school in June ranking toward the top of his class. Like so many blue-collar laborers, the man had opened the way for his son's part-time, entry-level job at the plant, but he hoped the boy would advance beyond the mill floor through education.

The steelworker advised, "Son, keep your nose clean, don't cause no ruckus and I know you'll move up the line. Who knows, someday you might even make foreman."

"Yes, Pa." The boy intended to work a full schedule after graduation while taking evening accounting courses three nights a week at the University of Pittsburgh. The father beamed with pride. His son might unlock the opportunity unavailable to him.

The second generation displayed the get-up-and-go and work ethic of their immigrant parents but spoke fluent English and possessed an American education. They showed up on time, worked hard and kept their mouths shut. Like their forefathers, they hated the sting of the workplace—heat, heat and more heat! Boiling steel lapped at the cauldrons, which hung precariously above the plant floor, threatening instant death. Noxious fumes and slippery surfaces pointed the roadway to loss of limb. Evening studies at local colleges provided a welcome respite from the stink of the mill.

Advancing age and hard labor had sucked the oomph from the old-timers. Muscle tears, cracked bones and stitches had taken a toll. Once powerful biceps strained from lifting. The first generation of steel veterans tired quickly. Some developed hacking coughs, crippling arthritis and failing hearts. These immigrant work hands rarely complained aloud to their families, but it hurt the sons that the bosses treated their fathers like worn-out mules after decades of loyalty. Those with a few dollars set aside might retire, but most plodded on and on.

Working conditions in the early twentieth century remained appalling. The younger men groaned to one another, "They treat us like crap. We got to look after ourselves, 'cause no one else is goin' to do it. We get piss-poor wages in this hellhole and work like dogs." After the Homestead killings crushed the AAISW, the company stifled unionization by ousting malcontents and troublemakers. The younger men talked big at the local bars over a glass of beer and a shot of whiskey. Changes were in the wind.

• • • • •

Carnegie Steel and H.C. Frick Coke had merged on April 1, 1900. With Henry Phipps fully retired and Frick ousted, Charles Schwab ran the show for Andy Carnegie. Charlie's huge salary and $19 million in stock (equivalent to $400 million in twenty-first-century dollars) made him rich. He purchased the six-acre Highmont estate in the plush east end of Pittsburgh from transportation baron Jacob Vandergriff as a tangible sign of his success. Schwab looked forward to the challenges of the new century, anticipating huge rewards as compensation for a big-league, high-stress job.

Charlie Schwab worked hard, never smoked and drank sparingly. However, he accumulated his share of vices: living large, a proclivity for taking chances, high-stakes gambling and, most of all, an eye for the ladies. He hobnobbed with the A-list fast crowd: Commodore George Dewey of Spanish-American War fame, authors Mark Twain and O'Henry and big-time spenders like "Diamond Jim" Brady and "Bet-A-Million" Gates. Charlie's wife, Rana, felt ill at ease with her husband's highfaluting cronies. She preferred to curl up at home with a good book. Since Rana proved unable to have children, Charlie encouraged her to raise orphaned nephew Carlton Wagner as her son.

When Rana's younger sister Minnie contracted typhoid fever, the Schwabs invited her to recuperate at their home. Rana summoned family friend Dr. Marshall Ward to tend to Minnie. Although he was twenty-five years her senior, Minnie and the doctor fell in love and eventually married. The beautiful redheaded nurse who accompanied Dr. Ward became Schwab's lover—the result, a baby girl.

The scandal of a bastard child in Victorian Pittsburgh would have ruined Schwab. Straitlaced Carnegie frowned at the slightest hint of carnal weakness and certainly would have fired Charlie if he had learned of a baby. The affair ended quietly, and Schwab confessed both his indiscretion and its result to Rana, begging her forgiveness and vowing to remain faithful, a promise he undoubtedly failed to keep. The child remained their hush-hush secret. Charlie would look after his daughter throughout his life. When the mother died, he increased his financial assistance and even took his daughter to Europe on vacation. While in Southampton, England, on a solo spring holiday in 1900, Schwab scheduled a poker game with Andrew Mellon but cancelled at the last minute due to an "appointment." The next morning, Mellon witnessed Schwab in a hansom seated beside a beautiful woman, quite likely the previous evening's "appointment."

• • • • •

Contention circled the steel industry like a murder of hungry crows nose-diving to seize the last scraps of food from an empty November cornfield. Sales at Jones and Laughlin had slowed to a snail's pace. Cutthroat competition eradicated the weak from the marketplace. Mighty Illinois Steel, a division of Federal Steel, shuttered its furnaces. Only a combination of strong overseas sales, expanded product lines and significant cost reductions deterred financial declines at Carnegie. In desperation, Federal and National Steel sliced their pricing and drew up plans for an integrated steel plant to maintain market share.

Carnegie wrote in *Iron Age*, "The consolidation of the iron and steel interests is a natural evolution."[168] He warned of his own predatory intentions: "Carnegie Steel leads the field, and we intend to continue to pursue our dominance." He opted for an offensive attack—an expansion in product offerings. "I wish to be on record as giving the straight tip—sell finished goods of which great quantities are used of uniform pattern."[169] Previously, Carnegie Steel had concentrated on raw materials, ingots, structural beams and armament plates. Now, he intended to expand his sphere of production. Carnegie's June 22 letter to the board delineated a strategy to punish J.P. Morgan's National Steel for initiating a price war: "A struggle is inevitable, and it is a question of survival of the fittest."[170]

Carnegie took immediate action. He undercut National's reduced pricing with the explanation: "The lowest price given has proved to be a high price at time of delivery in a falling market. When you want to catch a falling stone it won't do to follow it. You must cut under it."[171]

Carnegie Steel's strength in Lake Superior iron ore through its ownership in Oliver Mining provided a huge competitive advantage. His furnaces contained the latest technology, making use of coke, gas or electric fuel, whichever worked best.

Carnegie held all the tools, and he planned to use them. He ordered his engineers to draw up designs for a new plant in Conneaut, Ohio, to produce industrial tubing. He intended to corner that market just as he had done to rails and structural steel. Carnegie purchased seven eight-thousand-ton ore steamers to counterbalance Rockefeller's power in the Great Lakes shipping lanes. A planned Bessemer and Lake Erie Railroad extension to Connellsville's coke fields would reduce the oppressive freight rates offered by the major railroad lines. Carnegie Steel's overall synergy had built a potent commercial force, unassailable by any one of its rivals. However, with these overt efforts, Carnegie had provoked rival manufacturers, the Pennsylvania Railroad, John D. Rockefeller and, most importantly, J.P. Morgan.

Tremors from Carnegie's bellicose actions rattled the metal industry to the quick. One of J. Edgar Thomson's larger customers, American Steel Hoop, sliced its raw steel orders 80 percent in protest. Another, American Steel and Wire, canceled its billet contract. Andy Carnegie had fired the initial cannon volley in the "Great Steel War." No individual foe possessed the firepower to answer the salvo in kind.

An earthquake of threats and counter-threats had spewed through the steel and rail industries at Carnegie Steel's aggressive tactics. Conversation at the hallowed tables of the Duquesne Club predicted dire circumstances for Pittsburgh's smaller producers. One mini-magnate complained, "Someone's got to stop him."

A swarm of steel executives flocked to J.P. Morgan's Madison Avenue offices in search of a savior. The portly multimillionaire with the bulbous nose sat ensconced behind his enormous roll-top desk. He wore a trademark black banker's suit, starched white collar and gray ascot. A neatly barbered mustache imparted an aura of elegance to his overall austerity. The great man listened to the grumblings of the frantic emissaries, masking his thoughts. Morgan sucked in a healthy swag from his lit cigar and exhaled a ring of smoke to soothe his nerves, engrossed by the cloud swaying to and fro.

One competitor griped, "They say the new pipe plant will run $12 million. The bastard will sink us all."

Morgan had heard enough. He required no prognosis from a herd of sheep. He gloried in structure and balance. Carnegie had delivered chaos and disruption. "Ah, Andy, Andy, you are trying to demoralize railroads and finished goods just like you demoralized steel. This I refuse to tolerate," he barked to himself. The solution required order.

Morgan recalled Carnegie's failed attempt years earlier at inserting a wedge between the New York Central and the Pennsylvania Railroads to increase competition and reduce transportation costs. The gall of the Scot still stuck in his throat. His personal intercession with the rival railroads on his yacht had squashed that Carnegie nightmare then, and he would do it again. Competition reduced profits—anathema to order. Carnegie's pontifications on peace, labor's rights and Carnegie Steel's destiny chafed like a burr beneath a saddle, making every decent Republican sick to his stomach.

The little Napoleon had mocked Morgan's company, commenting to the press, "I think Federal the greatest concern the world ever saw for manufacturing stock certificates, but they will fail sadly in steel."[172] Morgan nearly retched as he chewed over his dealings with this insufferable buffoon. "Carnegie may control 29 percent of the country's steel output, but I'm not sure he fully understands with whom he is dealing," he ruminated. Morgan's Federal Steel produced 1,223,000 tons, far less than Carnegie's 2,290,000, but a significant producer worthy of respect, and he, like Carnegie, demanded respect.

One historian wrote that Carnegie was a man who "had no business ethics to hamper him. Might was right. There was no Sherman Law. There was no Interstate Commerce or Federal Trade Commission and no Clayton Act. Had Mr. Carnegie to encounter these brakes on business, he would have had many tumbles."[173] Absent the legal roadblocks encountered by future industrialists, Carnegie possessed the clout to bully his smaller rivals.

Only J. Pierpont Morgan or John D. Rockefeller possessed the wealth, power and moxie to stand against Carnegie, whose company generated nearly $40 million in profits. A linkage of the Lake Superior Mesabi iron ore mines to a Conneaut pipe plant spelled a near monopoly by Carnegie—a killer of profitability for smaller and weaker steel producers. Carnegie saw his actions as a natural outgrowth of vertical integration—an astute business move. Morgan saw it as a direct assault on an orderly market. He disliked this purveyor of tall tales—a small man who hedged words to suit his needs. "Enough," Morgan muttered. Eyeing another puff of smoke eddy above his head and disappear, the banker determined to swat the irritating gnat. If he could not crush him, he would buy Carnegie Steel.

In an early foray, Morgan attempted to maneuver private discussions with Charles Schwab, the Carnegie president. Schwab dutifully advised the board on November 6, 1900, of J.P. Morgan's request for a meeting on matters of "mutual concern." He agreed to present any substantive proposals to Andy Carnegie before action. After ending the morning's preliminary agenda, the board voted to authorize the Conneaut land purchase. Schwab and Carnegie's pipe plant was moving forward. The rumbles from the meeting generated a raft of consternation throughout Federal Steel. Morgan had to act quickly.

Months earlier, prior to his departure from Carnegie Steel, Chairman Henry Clay Frick had regaled members of the New York banking community with a tour of the Carnegie plants followed by a gala dinner at the Duquesne Club. Frick also led the visitors through his burglarproof steel vault at the Union Trust, a company on whose board he sat along with Henry Phipps and Andrew Mellon. J. Edward Simmons and Charles Steward, two of the bankers entertained in Pittsburgh, reciprocated with a black-tie dinner at New York's University Club on December 12, tagging Carnegie president Charles Schwab as the guest of honor and the evening's speaker.

J.P. Morgan.

A who's who of eighty millionaire financiers attended, including August Belmont, Jacob Schiff, rail tycoon E.H. Harriman and, most importantly, J.P. Morgan, who sat directly to the right of the guest of honor. Andrew Carnegie made a brief appearance but left early due to a conflicting engagement at the Pennsylvania Society. Henry Phipps misplaced his invitation and never showed. Only Judge James Reed and junior partner Albert Case represented Carnegie Steel for the entire evening.

Charles Schwab, the thirty-eight-year-old Carnegie Steel chief executive, stood at the podium and preached a vision of a profitable industry through consolidation to the choir of bankers. He envisioned a single interrelated steel mega-manufacturer rather than dozens of smaller, ineffective plants. Such an all-powerful steel producer would increase output, offer better control, provide technological efficiencies and decrease manufacturing and distribution costs. The speaker's melodious voice, rich smile, dramatic hand movements and lively wit gave life to a mammoth steel utopia during a forty-five-minute speech that was interrupted by frequent questions.

Morgan had listened attentively to the message of order and efficiency, his fingers tickling an unlit cigar. At its conclusion, Morgan congratulated

Schwab and shook his hand: "Wonderful speech, young man." The financier carefully ushered the steel executive by the arm to the privacy of a corner-window seat where he buttonholed him for more than thirty minutes to discuss in detail several points that caught his attention. Cognizant of the fact that Conneaut would undercut National's tube costs by ten dollars per ton, Morgan grimaced at the thought of the potential profit about to leak from his steel investment. Schwab's vision of consolidation led credence to his own dream of an all-encompassing steel leviathan. As a Morgan partner summarized, "Morgan had seen a new light."[174] Charlie Schwab might be the very man to execute his plan to rid the industry of the meddlesome Scot and reinstitute order to the industry in one single motion.

The dinner became "one of those events which direct the destiny of a nation," wrote social historian Frederick Allen.[175] Morgan reviewed the evening with confederate John Warne Gates: "Should we approach Clay Frick? He's a smart one." Morgan liked and respected Frick. Gates nixed the idea: "Too much bad blood. Better forget Mr. Frick for now."

Gates proposed a "chance" meeting with Schwab at the Bellevue Hotel in Philadelphia, which might give the parties an opportunity to expand the vision into a concrete plan. Unfortunately, Morgan became ill, postponing the get-together. A few days later, Gates arranged a private dinner with Schwab in New York, followed by a visit to Morgan's Madison Avenue mansion. Morgan partner Robert Bacon joined the discussion in the library, which lasted until three o'clock in the morning.

"Is Carnegie serious about this pipe plant? asked Morgan.

"Absolutely," Schwab confirmed. Carnegie Steel planned to move into finished goods—a direct attack on an orderly marketplace.

"Can he be stopped?" Charlie hesitated before echoing the only sensible alternative—only the merger of Carnegie Steel into one giant steel trust controlling the market would halt the Armageddon. Schwab explained his reasoning: "My arguments were mainly four—the economies that would result from consolidation, the improvement of the general business situation, the benefit to labor, and the steadying of the steel trade." Morgan approved wholeheartedly of a steel trust, but the million-dollar question remained: would Andrew Carnegie part with his company?

Morgan intoned, "Charlie, if Andy wants to sell, I'll buy. Go find his price."[176] "And oh yes," he added. "Compile a list of other companies that might fit into such an arrangement."

Schwab drew up his candidates. He excluded those that duplicated capacity or lacked the necessary synergy, presenting Morgan with his findings. He purposely omitted Jones and Laughlin, Oliver Steel and other players that would cost too much or failed to add significantly. Morgan separately commissioned one-time Carnegie engineer Julian Kennedy to develop a list of plants for inclusion. When Schwab and Kennedy's lists meshed, Morgan knew he had selected the right man.

Schwab liked Andy's wife, Louise, and the feeling was mutual. He called on Mrs. Carnegie at her West Fifty-First Street home in New York and recapped his talks with Morgan. "Do you feel Andy might be willing to sell?" He paused, allowing the question to take hold. Andy indeed had discussed retirement on numerous occasions with Louise, and she thought he would listen to a reasonable offer. In fact, she hoped her husband would retire, and she suggested a golf game as the perfect time to broach the subject.

The very next day, Carnegie and Schwab bundled up to ward off the January chill and headed for the links. Golf always put Andy in a fine frame of mind, especially when he won, and win he did. After the last hole, the two dined in Carnegie's private hilltop cottage above the club. While Carnegie sipped a scotch, Schwab rehashed his meeting with Morgan. Carnegie listened quietly but appeared receptive. Schwab ended with: "Andy, Mr. Morgan asked you just to name your price."

"Charlie, you've given me much to consider. Let me sleep on it tonight and call on me in the morning."

"Name your price, Andy. Just name your price," Schwab repeated.

Carnegie had passed the age of sixty-five, although he looked and felt much younger. For many years, he had involved himself with gift giving—libraries in Braddock and Dunfermline, church organs throughout the country, the Carnegie Museum in Pittsburgh and Carnegie Hall in New York. Morgan's money would provide fodder for other important philanthropies and the pursuit of world peace. The millions of dollars received would enable him to atone for the predatory years, a balm to a guilty conscience and the completion of the promise he had made to himself in his hotel quarters thirty years earlier. A dedication to the human good would become his reason for existence. He now intended to abide by the rule: "The man who dies rich dies disgraced."[177] A premium price for Carnegie Steel would provide the necessary backing for his charities. If Morgan wanted to ante up, he would walk away with a bushel of cash—more than he could spend in two lifetimes. Taking a stubby pencil between his fingers, he tallied the value of his empire and jotted on a piece of paper:

Capitalization of Carnegie Company bonds to be exchanged for bonds in new company:	$160,000,000
$160,000,000 in $1,000 shares of stock in Carnegie Company to be exchanged for $1,500 shares of stock in new company:	$240,000,000
Profit of past year and estimated profit for coming year:	$80,000,000
Total to be paid for Carnegie Company and all holdings:	$480,000,000

Carnegie demanded an additional kicker: his share, that of cousin Dod Lauder and that of sister-in-law Lucy Carnegie must be paid in 5 percent, first-mortgage bonds. The demanded figure included a rosy estimate of earnings for the upcoming year and equaled nearly double the price offered by the Moore consortium just a few months earlier.

When Schwab delivered the note to Morgan, the banker glanced at the total without emotion and coolly responded, "I accept this price."

A few days later, Carnegie reconsidered, but Schwab convinced him matters had progressed too far to renege. On Monday, February 4, twenty-nine Carnegie partners—all future millionaires—inked the merger papers along with a fond farewell letter to their old boss. Phipps lay sick in bed with bronchitis when he learned of the sale from Dr. Jaspar Garmany, who also served as Carnegie's physician. His accountant's mind tabulated a personal take exceeding $50 million. He rolled over, looked up at the physician and chirped, "Ain't Andy wonderful!"[178]

On February 25, 1901, J.P. Morgan formally announced the birth of U.S. Steel. The merged steel colossus concentrated nearly 1 percent of the country's total wealth. The pieces of the steel empire fell into place quickly as each of the ten targeted companies joined. Tight-fisted Judge Elbert Gary, who handled the merger legalities for Morgan, negotiated the corporate details with precision. When John Gates threatened to hold out for a higher price, Gary announced with take-it-or-leave-it finality, "I am going to leave this building in ten minutes. If by that time you've not accepted our offer, the matter will be closed. We will build our own wire plant."[179] Gates caved.

Throughout the negotiations, Morgan and Carnegie never spoke to each other. In typical gentlemanly fashion, Morgan phoned Carnegie and invited him to his offices to shake hands over the largest deal ever consummated in America. The prickly Carnegie, feeling slightly under the weather, countered, "It is the same distance from my house to the Morgan Twenty-Third Street offices as vice versa. You should come here." Morgan graciously consented and motored to the Carnegie home for a celebratory meeting. As he entered the spacious entryway, Morgan offered a firm handshake to his host and proclaimed, "I want to congratulate you on being the richest man in the world." Carnegie beamed and responded with his characteristic high-pitched lilt, "I am the happiest man in the world. I have unloaded the burden on your back, and I'm off to Europe to play."[180]

Carnegie's personal share of the take amounted to $225,639,000 in 5 percent gilt-edged bonds delivered to the Hudson Trust Company in Hoboken for safekeeping. This figure equated to as much as $159 billion in current dollars.

In addition to selling for top dollar, Carnegie believed J. Pierpont Morgan, the "Wizard of Wall Street," might have bitten off a larger mouthful than he could chew. Should the banker default, he might just recover his company for a fraction of its value. "Pierpont feels that he can do anything because he always got the best of the Jews on Wall Street. It takes a Yankee to beat a Jew, and it takes a Scot to beat a Yankee," Andy boasted.[181]

Details flowed smoothly for the first month. President Schwab appeared atop his game. On February 25, Phipps wrote him a laudatory note: "For the success which you've done so much to bring about I congratulate you most heartily."[182]

Great wealth also opened Phipps's philanthropic horizons. For years, brother-in-law John Walker had chided him for penny-pinching. Phipps admitted, "John, I wish the time would come when I could get away from the feeling that a penny is a penny."[183] Well, the time had come. Now, he had more money than he ever could need. Carnegie had forgiven Phipps for siding with Frick, and the two old pals hobnobbed in Europe while enjoying their good fortune.

• • • • •

The men at J. Edgar Thomson tried to make heads and tails out of the merger. Uncertainty rippled along the floor. Would the big boys from New York cut wages? Any reduction, whether it be pay or hours, would rain hunger on the poor souls who barely eked out a living. They knew conditions would get no better—just another dose of heat, steam, acid and accidents—different owners but the same pain. The veteran laborers above age fifty worried less about their own welfare than that of their families. They knew their days at the furnace would end shortly. One old-timer had felt the strength ebb from his body with each day. Just that night, a sharp pain shot from his jaw down his left arm. He did not know what that meant, but it could not be good. Well, with the mill closed for a week of retooling, the man intended to rest, fix the broken door at the house and, best of all, take his son fishing on the Monongahela River.

• • • • •

While the sale flowed toward finalization, Oliver Mining president Henry Oliver cruised in the West Indies. A lieutenant from Oliver Wire telegraphed him with a notification of the U.S. Steel mega-merger: "Suggest return home at once to protect interests." Oliver acted unconcerned. Besides, National had contracted to take all the surplus ore not purchased by Carnegie Steel.

When Oliver returned to the States, he called upon Carnegie at his New York mansion to find out where he stood, knowing about the Scotsman's intention to sail for Skibo shortly.

"Well, Andy, I understand you have sold out to Morgan." Carnegie nodded in the affirmative.

Oliver continued, "I understand you also sold your five-sixths interest in the Oliver Iron Mining Company."

"I did."

"What did you do with my one-sixth?"

"I did nothing," answered Carnegie.

"What about our agreement to sell or not sell together?" asked Oliver. The pact of January 5, 1899 signed by Carnegie president Schwab specifically stated, "It is hereby mutually covenanted and agreed that neither of said parties will sell or dispose of any part or all of their respective holding in the Oliver Mining Company without giving to the other party thirty days notice in writing of its intention to do so."[184]

"We have no such agreement," dissembled Carnegie, who clearly had violated the pact made by his president.

"Very well," continued Oliver without displaying a hint of anger. "Would you mind telling me what you got for your five-sixths shares so I can get the same rate for my one-sixth?"

"No, I promised Mr. Morgan that I would not tell."[185]

With the interview at an impasse, Carnegie confirmed his intention to sail to Liverpool on his way to Skibo in Scotland. Carnegie had stonewalled

Henry Oliver.

him, but Oliver smiled politely and took leave of his duplicitous partner.

Undeterred, Oliver telephoned former Pittsburgh and Western Railroad associate and friend Thomas King and asked him to book passage on the same ship as Carnegie. Oliver knew exactly how to bait Carnegie. King should "casually-on-purpose" hook up with Carnegie and joke with him about selling too cheaply—a surefire inducement for the Scot to brag about the actual price received for his Oliver Mining stock.

By the time the ship reached Liverpool, Oliver received a King cable announcing his success at outfoxing the Scottish chatterbox. Henry Oliver now knew the price Carnegie received. After conferring with his confederates, he asked the secretary of Oliver and Snyder Steel, the title holder of Oliver Mining's stock, to draft three identical letters to J.P. Morgan with three differing selling prices, two of which were higher than that paid to Carnegie—each letter to be placed in a separate sealed envelope marked in pencil, "1," "2" and "3."

"Tell my brother that I will present to Mr. Morgan whichever one of these letters I think he is in the mood to accept and that I will wire you tomorrow which one he has accepted," Oliver advised.

The meeting with Morgan went well, and Oliver cabled his office: "Found party very receptive, have nine-and-one-quarter tons," code for a price of $9,250,000 in preferred shares and an equal amount in common of U.S. Steel stock—a higher price than Carnegie received but a good deal for U.S. Steel nonetheless. Charlie Schwab later estimated the total value of the company ore properties at $700 million, and the Oliver Mining portion represented the largest part. After Oliver also sold his Monongahela Tin Plate and Oliver Wire companies, he bragged to Duquesne Club cronies Henry Phipps, B.F. Jones, Henry Clay Frick and Andrew Mellon: "I guess I outfoxed Andy this time. I can't wait 'til he finds how much I got."[186]

Morgan officially appointed Schwab president of U.S. Steel on April 16, 1901. He also named Henry Clay Frick to the board, but not Andrew Carnegie. Schwab objected, "This was not according to our understanding." "Son, the Frick name will add luster," countered Morgan. Frick agreed to refrain from attending board meetings since he owned 25 percent of competitor Union Steel. Schwab gulped deeply but accepted Morgan's fiat without further complaint.

President Schwab's $100,000 glittery salary remained but a fraction of his huge profit-sharing bonus. He stood atop the Gilded Age's highest echelon of the salaried rich. Brother-in-law Alva Dinkey and younger brother Joe shared in the financial cornucopia. Alva rose to the presidency of the Carnegie division of U.S. Steel, and Joe served as special assistant to the president. Charlie relished his celebrity status. He purchased an entire New York City block that once housed an orphans' asylum for $850,000. He cleared the land and began construction of a monument to his accomplishments—a dream castle. As head of the country's largest corporation, he would build one of the largest homes in the country.

U.S. Steel's first annual report, which found its way into the homes of nearly sixty thousand stockholders, brimmed with illustrations of sixty-three furnaces and numerous plants. Its net profit for 1902 reached $133 million, and its total revenues trailed the entire United States government exclusive of the post office by a mere $2 million. It controlled 50 percent of the country's steelmaking capacity, and Schwab ran the show, or at least he thought he did. He promoted the mega-corporation's advantages to the media, the government, stockholders and the public: a single New York home office, tight control of ore supplies, orderly production, friendly inter-plant rivalry, a highly compensated labor force, freight savings and a free flow of information.

However, the stress of aligning the corporate jigsaw puzzle proved onerous. Schwab spent most of his energy soothing the egos of U.S. Steel's twenty-four-member board or quieting disagreements between the executives of the merged companies rather than concentrating on his strong point: steel manufacturing. By December, internal politicking had sapped his energy. He found himself isolated and alone. The savvy board of financiers pursued price stabilization, steady shareholder profits and their own personal agendas. Board member Marshall Field hyped a plant for his home city, Chicago. John D. Rockefeller pushed U.S. Steel to purchase his decrepit Troy, New York steel plant. Nearly all opposed Schwab's "excessive" modernization expenditures.

Judge Elbert Gary, the former president of Federal Steel, became board chairman and headed the all-powerful finance committee. He counterbalanced Schwab's extensive requests for investments in new equipment. The strong-willed executives rubbed against each other like sandpaper on silk. Gary reined in Schwab's plans for technological advancements more effectively than Henry Phipps and George Lauder had done at Carnegie. Schwab's grand schemes turned off the numbers-oriented Gary, who preferred price-fixing and order-sharing agreements to price competition and plant modernization. Gary believed dividends and pre-payment on notes due trumped the reduction of variable production costs. In short, steel was merely a byproduct to moneymaking. Even Charlie's fun-loving style grated on the puritanical judge, a prig who frowned on gambling, loose living and liquor. At Carnegie, each attending director had received a twenty-dollar gold piece. To lighten pressure, the directors flipped coins to win the honorarium of any missing directors. Gary, a staunch Methodist, nixed such sinful foolishness. Schwab carped, "Judge Gary had no real knowledge of the steel business, forever opposed me on some of the methods and principles that I had worked out with Carnegie—methods that made the Carnegie company the most successful in the world. I have been hampered, criticized and goaded by incompetent critics, who do not understand the whole steel situation."[187]

Schwab's tendency to share his opinions in public offended many. His pronouncement to the press that "politicians who attempt to obstruct industrial development are attempting to obstruct human progress" irritated Congress. His statement that work experience outweighed an education created a storm among the college intelligentsia. Snippets that unions "narrow opportunity, dull ambition, and give no man a chance to rise" set off the hackles of journalist William Randolph Hearst, whose liberal newspapers unleashed a stream of protest. When Schwab confessed to selling abroad at reduced prices, United States customers accused him of overcharging. The press labeled his argument that the strategy "assured full capacity" and benefited the American worker as "illogical gall."[188]

The public also chafed at his outrageous personal spending, symbolized by his recently built $3 million, seventy-five-room Riverside Drive palace modeled after Chenonceaux, a Loire château. The house contained its own power plant, which burned nearly ten tons of coal daily during the winter months. Spectacular decorations included a $100,000 Aolian pipe organ, a Henri IV paneled library and a $200,000 Louis XIV dining room highlighted by artist Jose Villagos's ceiling mural, *The Prosperity of*

America. The 60-foot swimming pool, private gymnasium, bowling alley, six elevators and 116-foot tower all epitomized the outrageous consumption of the Gilded Age. Even Carnegie wrote, "Have you seen Charlie's house? Mine is a cottage by comparison."[189] Meanwhile, many of Pittsburgh's blue-collar laborers lived on unpaved streets in crowded smoke-tinted row houses lacking indoor plumbing and central heating.

While at Carnegie Steel, Andy had maintained a tight leash on his young executives, saving them from "wine, wagers and women."[190] Schwab had concealed his wayward behavior from the boss. He drank moderately—an occasional glass of champagne or a sip of brandy or scotch. He never smoked. He liked good food, adding a pound or two with each passing year. He was anything but a clothes horse. In fact, when a coworker learned Schwab had hired a valet, he burst into laughter. "What for? He only has one suit."[191]

However, Charlie's taste for the gambling tables, expansive lifestyle and eye for the ladies offended Judge Gary. Schwab frequently visited the casinos in Monte Carlo. On December 26, 1901, sapped by the pressure from an intrusive board and a full schedule, he sailed to Europe on the liner *La Savoie* accompanied by the family of Charles Schoen, the founder of Pressed Steel Car Company and a partner of Henry Oliver in the manufacture of metal railway cars. On January 8, Schwab motored to Monte Carlo absent his wife in a sleek roadster. It was his fortieth birthday. He intended to enjoy a night of fun and gambling.

A reporter witnessed Schwab hit a lucky streak. Sniffing out a feature story, the *New York Sun* headline blared "Schwab Breaks the Bank" after the U.S. Steel president won $7,500 by placing maximum table bets in an era when a steelworker might earn less than $400 for a full year's labor. High-stakes gambling stood in direct opposition to the conservative morals of the time. The heavy betting shocked Carnegie, who wired his former protégé, "Probably have to resign. Serves you right." Carnegie followed up his cable with another to Morgan calling for Schwab's resignation: "He is unfit to be the head of the United States Steel Company—brilliant as his talents are—I recommended him undeservedly to you. He shows a sad lack of solid qualities, of good sense, and his influence upon the many thousands of young men who naturally look to him will prove pernicious to the extreme."[192]

Schwab assumed the negative media publicity and his mentor's barbs had sunk his career at U.S. Steel. A thoroughly distraught and chastised Schwab penned a letter to Carnegie before traveling home: "I admit I have made a serious mistake and one I shall probably never be able to rectify, and I will pay the penalty. I have cabled Mr. Morgan again today saying that he must accept my resignation. My chief pang is not for my loss of position but the loss of your confidence and friendship."[193]

• • • • •

A plant closing for retooling allowed steelworkers a few days to refuel their tanks. Time off provided one old-timer with a chance for rest and recuperation. He had walked to the library on Monday and borrowed another Jules Verne book, *Around the World in Eighty Days*—a trip through fantasyland. He fixed the creaky front door at the house and enjoyed what he labeled a good brisk sit.

Although the worker had paid off the house—all $800—retirement seemed out of the question. He had not saved enough. No matter how hard he worked, he never saved much, and he was one of the lucky ones—a skilled furnace man.

All week he had looked forward to fishing with his son on Thursday evening on the Mon River, and the experience did not disappoint him. The guys borrowed poles from a neighbor and distant cousin. The son somehow obtained worms, and the pair meandered to the river. Only one or two hardy souls sat on the bank, braving the chilly air.

"You think this is good idea, Papa? I'd hate to see you get sick."

"Yea, it's good idea. We get chance to talk," replied the father. "How's work?"

"Fine, Papa."

"And school?"

"I am taking bookkeeping, and it's fine, too."

"You learn much?"

"Yes, Papa. The foreman talked to me about a training program at work—maybe next year."

"Good, I want to see you move ahead. All my life I big man and strong man. You big, too, but you use head instead of arms. All my life I work for little men, but they have big dreams and move ahead. You have big dreams. You make me proud."

"Yes, Papa," answered the son, and the two men sat quietly and fished without success, but the father enjoyed every moment.

• • • • •

Schwab returned to New York, certain his tenure at U.S. Steel had ended. Although the high-stakes gambling scandalized dour Judge Gary, Charlie's "bully good time" had not bothered Morgan, who was known for his own amorous escapades and sensual appetite.[194] He refused to accept the letter of resignation. A grateful Schwab wrote Morgan partner George Perkins, "I'll do anything Mr. Morgan wants. He's my idea of a great man. Carnegie has condemned me without a hearing. Mr. Morgan, a new friend, is broader by far. I'm his to command."[195]

Schwab backed Morgan in every sense. When Pittsburgh native George Westinghouse, a man he liked and respected, refused a proffered merger with Morgan's General Electric, Schwab denigrated the inventor's business savvy to business reporter Clarence Barron in an interview: "George Westinghouse lived and slept air brake, but when it came to other things he could not give them his attention to detail. Had he devoted himself the same way to Westinghouse Electric, he would have made the same success, but a man cannot follow the details of many things."[196]

To assuage his conscience and polish his tarnished image, Schwab donated $65,000 for a chapel at State College in Centre County. He sent $2,000 for the poor in Braddock and completed his pledge for a new church at Loretto. Carnegie, whose shock cooled over his protégé's indiscreet behavior, chipped in with $8,000 for the church organ. Schwab vowed to watch his ways and moderate his treatment of the workforce—so long as he could maintain profits and production.

Job pressure, board criticism and the media assault had built up a reservoir of psychological and physical stress. Schwab shed pounds and fell prey to depression. His legs had grown numb, and he suffered from insomnia and acute anxiety. His physician prescribed rest. On August 20, 1902, Schwab again offered his resignation to Morgan, who countered, "Charlie, take some time off. You'll be fine when you return." The very next day, Schwab departed by ship for Aix-les-Bain in France. A letter from Louise and Andrew Carnegie in his stateroom wished him a speedy recovery.

In France, a bogus newspaper article announced Schwab lay on his deathbed and wished to give away his fortune to the needy. Hordes of money-hungry adventurers hounded him, forcing a hasty departure to England. Through all his misfortunes, Schwab managed to maintain a small coterie of defenders. One newspaper labeled the "planted story" of his imminent demise as reprehensible. Friends hinted that Frick might have engineered the board's constant opposition as retribution. Other cronies blamed Judge Gary.

Important decisions proceeded without Schwab's input. In December 1902, U.S. Steel purchased Union Steel from Frick, Mellon and Donner while he recuperated abroad. With the Union Steel conflict-of-interest issue laid to rest, Frick accepted an active board role, Judge Gary appointing him to the all-powerful finance committee—nicknamed the "Crown Ring." Clearly, Schwab's fortunes at U.S. Steel had declined.

Charles Schwab returned to the States on March 17, 1903, but remained homebound to complete his recuperation while supervising the $21,500 expansion of his Aolian organ. The company continued to function without his active participation until early spring, when he advised the board, "I have come back with new health and vigor."[197] Schwab presided over the April 20 annual meeting at Hoboken, and the board reelected him as president, with William Corey serving as his assistant to ease job pressure.

• • • • •

Many of the J. Edgar Thomson veterans from the mid-1870s had already died or retired. One fifty-plus worker awoke with a start. His chest pounded. He felt like one hundred pounds of bricks pushed against him. His left arm tingled, and his head ached. He had lost all his strength and could not rise from the bed. He had rarely missed work, and then only due to catastrophes like the mill accident that scarred his face and almost killed him and the day of his wife's funeral. He had even worked the day his wife had delivered their son at home with the help of a midwife.

Frightening thoughts poured through his mind. Could this be the end? Would he live long enough to witness his son's twentieth birthday? He sucked in a deep breath with effort and exhaled—deep breath and exhale—deep breath and exhale. It helped. The pain passed and clarity returned—maybe just the onset of the flu. He got up from his bed and dressed. No sense missing work. He needed the pay.

• • • • •

With his presidency intact, Schwab attended the ribbon cutting for the C.M. Schwab Free Industrial School in Homestead, to which he had donated $200,000. While in Pittsburgh, he visited both plants he had supervised. Homestead had looked shipshape, and he expected the same from J. Edgar Thomson. Schwab generally identified a line worker or two on the floor for conversation. He watched a muscular youth push a tram across the room. "Keep up the good work," he prodded.

As he passed Furnace D, he came across an older man who looked familiar.

"You been here a long time?" Schwab asked.

"Yes, sir, almost since the day she opened." The worker instantly recognized his old superintendent Charlie Schwab.

"You look a little ragged. You okay?"

"Just a little under the weather—nothing to speak of."

"Well, keep up the good work."

The old laborer steeled up his courage and put in a good word for his boy. "Thank you, sir. My son just started working full time at the plant, and he's taking college classes at the university evenings."

"Sounds like a go-getter."

"Yes sir, he a hard worker."

Schwab's mind raced back to the time he had sold Captain Jones a cigar at the general store. The superintendent had taken him under his wing. There was something he liked about this old coot. Everyone could use a boost. Things had been going poorly for him lately, but maybe he could do something for the young man.

"Where's your son?"

The steelworker pointed across the floor to a dirty-faced muscular youth. Schwab approached a teenager diligently shoveling coke.

"Young man."

The youth looked up at the distinguished-looking executive dressed in a suit.

"Your dad says you are going to college at night. What do you want to do with your life?

"I love steel, sir. I want to make steel."

"And so do I. That is just the very answer I would have made at your age."

"Thank you, sir."

"Do you wish to use your head or your hands?"

"Most certainly my head—or my head and my hands if necessary, but I hope to move ahead."

"And so you shall if you apply yourself. Report to personnel tomorrow morning. We'll see what we can do."

Schwab felt good as he left the plant. He was a German Catholic of poor parentage in a Scottish Presbyterian world. The steelworker definitely was not Scottish and not Presbyterian. Just possibly, he might have changed a life.

Pittsburgh—city of steel.

The weary old man finished the day with a warm feeling in his chest. He had helped groom his son's future. Had he stayed in bed, he never would have spoken with Mr. Schwab. He slowly moved homeward on that cool late winter night with a smile on his face. He neared the age of sixty, an advanced age for a steelworker, much older than his father or grandfather had lived. He had confidence that his son, who went to class after work, would be fine. He had met all the hell life had thrown at him face to face and still retained dignity and a faith in God. As he looked up at the horizon, he could see the moon—a bright shiny moon welcoming him.

Tiny snowflakes caressed his cheeks and pelted his shoulders. Yes, God was good. Then, it struck without warning. A thunderbolt slammed against his chest as he fell to the cold ground. He could not breathe. Pain struck with violence. This would be his end. He would die alone and without last rites. As he gave up any hope for a final redemption, a vision of his dear departed wife appeared. He felt soft lips caress his brow. He thought he heard her whisper sweetly, "Be not afraid, my dear," and he wasn't. He would not die alone without final rites. His wife was with him. His eyes closed as his body shut down. He was gone—hopefully to a better place.

EPILOGUE

A chain of unfortunate circumstances ended Schwab's career at U.S. Steel. Less than eight weeks after the Carnegie merger, Schwab acquired controlling interest in Bethlehem Steel, the country's second largest steel-plate maker. Upon reconsideration, he thought better of the purchase. Worried his board would accuse him of a conflict of interest, he sold his shares in Bethlehem to U.S. Steel for a small profit.

In May 1902, naval architect Lewis Nixon requested a loan of $500,000 from Schwab to shore up his faltering company, U.S. Shipbuilding, a merger of seven shipyards. In exchange, Nixon promised to purchase all his steel requirements from U.S. Steel. During further negotiations over a lunch at the Lawyer's Club in New York on June 12, Nixon proposed that U.S. Shipbuilding purchase Bethlehem Steel as well.

Schwab sought a cash price of $9 million, a sum Nixon could not raise. Schwab countered with a figure of $10 million in stock and $20 million in bonds in lieu of cash. Nixon surprisingly expressed interest. In a complex swap, Schwab reacquired his Bethlehem stock from U.S. Steel for $7,245,000 and shifted the shares into the U.S. Shipbuilding coffers. U.S. Shipbuilding next guaranteed Schwab sufficient dividends to cover a 6 percent interest rate on Schwab's investment and guaranteed Bethlehem's plant and equipment as security against default. In a final move, Schwab transferred $2,500,000 of his own stock to U.S. Steel as their profit on the transaction. Schwab gloated. He controlled nearly $27,500,000 of U.S. Shipbuilding's $71 million in capital. The deal appeared shrewd, but not for long.

U.S. Shipbuilding's sales skidded, scuttling its profitability, although subsidiary Bethlehem Steel continued to prosper. With bankruptcy threatening, Nixon petitioned Schwab for an immediate influx of capital from subsidiary Bethlehem to prop up finances. Schwab refused. With Nixon pleading for help to stave off receivership, Schwab released $250,000 from Bethlehem—just enough to pay the bond interest owed him.

U.S. Shipbuilding continued in a steep decline. The company needed an immediate bailout. Schwab offered to provide $2 million in exchange for the primary lien on all assets as collateral. A vocal contingent of stockholders denounced Schwab's harsh terms, forcing U.S. Shipbuilding into receivership on June 30, 1903.

A *New York Evening Post* editorial damned Schwab for his greed: "A man who would ostentatiously stake thousands on the red might make ducks and drakes of their property by transferring the scene of his gambling operations from Monte Carlo to Wall Street."[198]

The depth of the U.S. Shipbuilding scandal, public uproar concerning his over-the-top home, high-stakes gambling and constant bickering with Judge Gary pushed Schwab's resignation from U.S. Steel at the August 4, 1903 board meeting. "I am suffering from nervous trouble and threatened with a complete breakdown," bemoaned Schwab. Morgan allowed him to leave with dignity, praising him publicly for his "unequaled powers as an expert in the manufacture of steel."[199] Judge Gary had eliminated a troublesome burr and replaced him with another ex–Carnegie executive as president, William Corey. However, Gary controlled the purse strings. Corey eventually would fall prey to the same fate as Schwab—an affair with a showgirl and a divorce would lead to his forced resignation eight years later.

• • • • •

Ben Franklin Jones retired from full-time management of Jones and Laughlin at the turn of the century, shifting responsibility to his son. Under the control of the elder B.F., the company had grown into one of the leading steel manufactures in the country. The young bucks of Pittsburgh lauded B.F. as the leading pioneer of his age. With time on his hands, he dined with old pals like Henry Oliver at the Duquesne Club, rehashing the good old days.

Jones had played a starring role in the iron and steel industries during his past forty-five years in business. His sliding-scale wage scheme held labor tensions in check during the early years. He built one of the first distribution centers in the city. He championed river transportation and later invested in railroad delivery systems for both raw materials and finished goods.

Jones had witnessed the rapid growth of big steel in Pittsburgh. He watched coke replace wood and charcoal as the primary fuel and later instituted water, steam, gas and electric power. The Bessemer technology of the final quarter of the nineteenth century had relegated puddling to the scrap heap. Now, the modern open hearth was squeezing out the Bessemer furnace. Iron and steel virtually eliminated the use of commercial wooden bridges. Metal ships, skyscrapers and machinery fed the country's modernization into the twentieth century. B.F. had seen it all.

Throughout his life, Jones possessed a quiet strength that was respected by his contemporaries. He remained an active participant on the Jones and Laughlin board until his death in 1903 from neuralgia at the age of seventy-eight. Andrew Carnegie eulogized his friend to the *Pittsburgh Post*: "Benjamin Franklin Jones, the Nestor in manufacturing, is gone." Jessie Marian Isaacs's portrait of Jones hangs on the wall in the Founders' Room at the Duquesne Club. Although son Ben Franklin Jones Jr. would grow the company to an even higher plateau, Jones and Laughlin would fail along with many other steel companies in the late twentieth century due to international competition, high labor costs, unionism and a failure to keep pace with modern technology. Barely a hint of B.F. Jones's fame would survive into the twenty-first century.

• • • • •

With the funds he received from J.P. Morgan, Henry Oliver bought into a string of lucrative investments: the Standard Steel Car Company of Butler, the Chemung Iron Company, the Bisbee Copper Mine (a predecessor to Phelps Dodge) and $12 million in prime downtown Pittsburgh real estate. The Duquesne Club bestowed the honorary title "Father" on him for his years of service. This consummate diplomat had prospered in a rough-and-tumble age by taking huge risks in exchange for even larger rewards. He bounced between success and failure until he struck it rich with Oliver Mining. His knack for digging his way through an abyss of disaster and crawling to the top with precious gems overflowing from his hands amazed his contemporaries. Oliver's broad smile and win-win attitude overcame adversity. The word "can't" rarely entered his vocabulary. He relied on inherent Scotch-Irish optimism to lift him to the top like cream on milk. An ounce of luck, perfect timing and the gift of the right word at the right time located investors and formed strong relationships.

Henry Oliver died at home from nephritis following a six-week illness on February 9, 1904, with an estate of $45 million, worth more than $1 billion in current dollars. Friends R.B. Mellon and Henry Clay Frick pushed the *Pittsburgh Press*'s endorsement to rename Virgin Alley as Oliver Place, a fitting remembrance to one who had done so much for Pittsburgh. Like his pal B.F. Jones, Oliver's portrait by William Thorne hangs on the walls of his beloved Duquesne Club.

• • • • •

Seventy-five-year-old J. Pierpont Morgan died on March 31, 1913. He had founded General Electric and U.S. Steel, and his bank controlled dozens of future Fortune 500 companies, such as International Harvester and the Chesapeake and Ohio Railroad. He single-handedly led the charge that saved the country from financial disaster during the Panic of 1907. Following his death, Henry Clay Frick acquired several major art masterpieces from his renowned collection highlighted, by Jean-Honoré Fragonard's mural *The Progress of Love*, today displayed in New York's Frick Museum.

• • • • •

With his retirement from business, Andrew Carnegie immersed himself in gift giving, writing and the pursuit of world peace. Undoubtedly, he proved egocentric to the bone—unethical when it suited his needs and often unfeeling. On the other hand, he doled out contributions like no other philanthropist before him. He founded the Carnegie Corporation, the Carnegie Hero Fund, the Carnegie Foundation for the Advancement of Teaching, the Church Peace Union and the Palace of Justice at The Hague; underwrote 2,811 libraries; and donated 7,689 church organs. The onset of World War I crushed the optimism of the Scottish pacifist, who never again would visit his beloved home in Scotland. Andrew Carnegie died on August 11, 1919, at age eighty-three.

• • • • •

Henry Clay Frick continued his business career after leaving Carnegie Steel. He co-founded the Union Trust in Pittsburgh with crony Andrew Mellon and constructed the one-thousand-room Art Deco William Penn Hotel. He became the country's largest railroad investor and erected a monumental mansion on New York's West Side that eventually became the Frick Museum.

Frick gifted charitable causes without fanfare, such as founding the Henry C. Frick Educational Commission for the Enrichment of Teachers. When the Dime Savings Bank failed, robbing forty thousand children of their Christmas savings, Frick anted $170,000 to cover the losses of the juvenile depositors, making him a real Santa Claus.

Frick carried his hatred for Andrew Carnegie to the grave. A critically ill Carnegie tried to make peace with his adversary. Carnegie instructed his secretary, James Bridge, to hand deliver a conciliatory letter to Frick at his home. "So Carnegie wants to meet me, does he?" Frick muttered as Bridge read the letter aloud before handing it to him. Such a letter from a dying man might have evoked a hint of forgiveness in many, but not Frick. "Yes, you can tell Carnegie I'll meet him." Frick wadded the letter into a ball and handed it back to Bridge. "Tell him I'll see him in Hell, where we both are going."[200]

Frick died on December 1, 1919, at age sixty-nine, less than four months after Carnegie. He added fine art to his collection until the end. Frick's final purchase, *Mistress and Maid* by Jan Vermeer, occurred just months before his passing. The *Pittsburgh Press* called him "an old curmudgeon with a clunk of iron for a heart."[201] He was a hard man to pigeonhole. He spoke few words but said much. He possessed a brilliant mind and hated unionism. He based decisions on facts and figures—rarely sentiment. On the other hand, he loved his family and proved a brilliant businessman, a discerning art collector and an important contributor whose $117 million of donations dwarfed those of other important Pittsburgh gift givers like B.F. Jones, Henry Heinz and George Westinghouse. His organizational skill, strength of conviction and focus built up a host of admirers. His ruthlessness brought enemies. When Alexander Berkman, about to be deported to Russia as a punishment for communist agitation, learned of Frick's death, he sarcastically retorted, "Deported by God! It's too bad he cannot take the millions amassed by exploiting labor with him."[202]

As a final cruel joke, the members of the Duquesne Club hung Frick's portrait directly across from his archenemy Andrew Carnegie, locking them together for eternity.

• • • • •

Judge Gary and Schwab eventually patched up their many differences. In fact, Schwab presented Gary with a gold vase at a dinner party held in his honor where he conceded that although the two had disagreed, "I was wrong in most instances."[203] Judge Gary would run the U.S. Steel board with an iron hand until his death at age eighty-two on August 15, 1927.

• • • • •

Flush with cash from the sale of the century, Phipps buried the hatchet with old pal Andy Carnegie and joined the ranks of significant philanthropists. He underwrote worker housing in Pittsburgh and New York with amenities including individual bathrooms, steam heat and gas stoves—all at affordable rents. He donated more than $350,000 in aid for South Africans dislocated during the Boer War. He funded the Henry Phipps Psychiatric Clinic of Johns Hopkins University in Baltimore and supported research for the cure of tuberculosis in Philadelphia.

Carnegie mocked Phipps's soft heart for the sick and downtrodden in lieu of bricks-and-mortar causes: "You just can't make Henry see what an unwise sinner he is. I've many a year tried to be his guardian, but he won't be reasonable on this one special weakness that keeps him spilling largess around—not bettering but spoiling humanity, being a helper as he calls it—being a hurter as he ought to call it, paying out a fortune every year to those who didn't earn it."[204]

Phipps, like Frick, had dumped his watered-down U.S. Steel stock near its high after the merger, repurchasing it when it dropped. He wisely invested in downtown Pittsburgh real estate and, like John D. Rockefeller partner Henry Flagler, invested in prime South Florida beach property, adding millions to his immense wealth. He served on the Mellon, Union Trust and U.S. Steel boards with Frick.

Old age treated Phipps poorly. By 1914, he had resigned from most business interests. His body stooped; his mind deteriorated. While sitting in the Waldorf Astoria lobby, a large jovial man greeted him warmly with his hand outstretched in brotherly fashion. A confused Phipps mumbled, "I'm afraid I don't recall you." Former United States president William Taft shrugged his shoulders, smiled at the frail old man and ambled down the hallway.[205]

Phipps spent his final decade in illness and seclusion. Death finally released ninety-year-old Henry Phipps on September 22, 1930. His final estate, once valued at more than $100 million (close to $2,500,000,000 in current dollars), amounted to only $2,913,805. He had gifted the bulk away to family and charity.

• • • • •

Andrew Mellon, as lead officer in Mellon Bank, played a key role in the founding and growth of Alcoa, Union Steel, Koppers, Carborundum and Pittsburgh Coal. He served as the secretary of the treasury under Presidents Harding, Coolidge and Hoover and once paid the third-largest taxes after John D. Rockefeller and Henry Ford. He died on August 26, 1937, bequeathing the bulk of his art collection to the Smithsonian Institution. Helen Clay Frick, daughter of his friend, wrote fittingly, "The world does not make such men any more as these fathers of ours."[206]

• • • • •

The resignation from U.S. Steel humiliated Schwab, but Bethlehem Steel provided him with a new lease on life, largely at the expense of the U.S. Shipbuilding shareholders. He hired top-notch management professionals and updated the plant's technology. He bragged, "When I organized Bethlehem, I took all the leftovers and made them the best."[207] The newly patented wide-flanged H-beam provided Bethlehem with the means to revolutionize the building of skyscrapers. The hiring of top-notch executives like Eugene Grace supplied the necessary leadership. Carnegie crowed about Charlie: "Mr. Schwab is a genius. I have never met his equal."[208]

The submarines Schwab built for England during World War I may have swayed the outcome of the conflict, but Schwab could be ruthless toward his workforce in achieving his ends. A Bethlehem skilled machinist earned fifteen to twenty-seven cents per hour for a six-day, sixty-two-and-a-half-hour week. Unskilled labor earned even less—a dismal twelve and a half cents per hour for up to fourteen-hour days.

Bethlehem allowed for Charlie's return to the game. In 1917, he wrote the book *Succeeding with What You Have*, a collection of conventional wisdom that included morals like success comes from "working a little harder than anyone on the job."[209]

When the United States entered World War I, the government conscripted Schwab as director general of the Emergency Fleet Operation. The *Literary Digest* wrote, "The ships will be built. Mr. Schwab knows steel, knows ships, knows how to handle labor, and has a prestige which carries weight with every businessman in this country and abroad."[210] Schwab goaded, prodded and pushed production. Thomas Edison called him a "master hustler." He had become a hero.

The Depression sunk both Schwab's health and fortune. Loans to family and friends and sour investments in a dining car company, a silk mill and a zinc smelter sucked his net worth. At his brother-in-law Alva Dinkey's funeral, he confessed to another attending steel executive, "I am very much ashamed of the way we treated our labor."[211]

By the end of 1933, only eight of the fifty-one Carnegie Steel executive veterans remained alive. Schwab recalled the sad words of William Dickson commemorating the loss of his brother Joe some years earlier: "When I remember all the friends so linked together—I've seen around me fall like leaves in wintry weather, I feel like one who treads alone."[212] Seventy-eight-year-old Charlie Schwab died from a heart attack and complications of diabetes on September 19, 1939. The liabilities of his estate exceeded the assets by $338,349. He joked about the difference between him and Carnegie on wealth distribution: "I spent mine. Spending creates more wealth for everybody."[213]

Looking back on his life, he deplored his lack of sensitivity and his unfaithfulness to his wife. He sadly mused, "It hurts me very much to be branded as nothing but a selfish, self-seeking, mercenary, merciless fellow, callous toward workman and toward everyone else."[214]

With the passing of Schwab, the age of steel in Pittsburgh had peaked. World War II generated a temporary spurt, but the 1980s signaled the death knell for giant steel throughout the country. Jones and Laughlin and Bethlehem Steel went out of business. China and Japan outproduced the United States, with India, Russia, South Korea, Ukraine, Germany and Brazil trailing close behind. Currently, the United States manufactures less than 6 percent of the world's total steel. Today, we remember the steel titans primarily by the monuments to their names: the Phipps Conservatory, the B.F. Jones Library in Aliquippa, Schwab Hall at St. Francis University, Carnegie-Mellon University, the Oliver Building and Frick Park.

• • • • •

The first wave of immigrant steelworkers enriched Pittsburgh's history, although most were buried in plain pine coffins. Some only could sign their names with an X. Most left the earth with minimal possessions—but perhaps more than their one-time boss Charles Schwab: possibly a

Steel mill.

small row house, some furniture and a few dollars after covering burial expenses. These steelworkers served as a vital cog in the growth of our country. Their hard work enabled Carnegie and later U.S. Steel to become the world's most powerful steel company. Their children would join the union and earn their way to the middle class. Most received an education through the scrimping and saving of their parents. The smartest or luckiest advanced up the ladder as doctors, lawyers, accountants, retailers or middle management in the steel industry. These second- and third-generation Americans would marry well, raise their own children and make this country strong, but that is another story.

NOTES

Chapter 1

1. Lubove, *Pittsburgh*, 9–10.
2. Bridge, *Inside History*, 194.
3. Warren, *Triumphant Capitalism*, 51.
4. Zubrinic, "Joe Magarac."
5. Samuel, *Rich*, 10.
6. Lorant, *Pittsburgh*, 163.
7. Herman, *How the Scots Invented the Modern World*, 386.
8. Lloyd, *History of the Jones & Laughlin Steel Corporation*, 4.
9. Wollman and Inman, *Portraits in Steel*, 19.
10. Childs, *Life and Times of John Pitcairn*, 130.
11. Davis, *Phipps Family*, 14.
12. Boegner and Gachot, *Halcyon Days*, 13.
13. Ibid., 14.
14. Wall, *Andrew Carnegie*, 995–96.
15. Ibid., 84.
16. Dietrich, "Andrew Carnegie," 115.
17. American Experience, "People and Events: Andrew Carnegie."
18. Dietrich, "Andrew Carnegie," 116.
19. Nasaw, *Andrew Carnegie*, 38.
20. Livesay, *Andrew Carnegie*, 24.
21. Krass, *Carnegie*, 65.

22. Swetnam and Smith, *Carnegie Nobody Knows*, 22.
23. Carnegie, *Autobiography*, 85.
24. Ibid., 88.
25. Wall, *Andrew Carnegie*, 244.
26. Ibid.
27. Bridge, *Inside History*, 11.
28. Ibid., 17–18.
29. Krass, *Carnegie*, 78.
30. Nasaw, *Andrew Carnegie*, 94.
31. Krass, *Carnegie*, 103.
32. Livesay, *Andrew Carnegie*, 70.
33. Carnegie, *Autobiography*, 129.

Chapter 2

34. Casson, *Romance of Steel*, 87.
35. Misa, *Nation of Steel*, 23.
36. Mellon, *Thomas Mellon and His Times*, 24.
37. Livesay, *Andrew Carnegie*, 96.
38. Standiford, *Meet You in Hell*, 46.
39. Chernow, *House of Morgan*, 39.
40. Livesay, *Andrew Carnegie*, 100.
41. Casson, *Romance of Steel*, 127.
42. Livesay, *Andrew Carnegie*, 100.
43. Krass, *Carnegie*, 165.
44. Burnley, *Millionaires and Kings of Enterprise*, 55.
45. Hessen, *Steel Titan*, 10.
46. Warren, *Industrial Genius*, 4.
47. Hessen, *Steel Titan*, 16.
48. Ibid., 20–21.
49. Warren, *Industrial Genius*, 7.
50. Hessen, *Steel Titan*, 18.
51. Misa, *Nation of Steel*, 29.
52. Casson, *Romance of Steel*, 91.
53. Warren, *Industrial Genius*, 8.
54. Hendrick and Henderson, *Louise Whitfield Carnegie*, 53.
55. Ibid., 54.
56. Ibid., 59.

57. Livesay, *Andrew Carnegie*, 127.
58. Nasaw, *Andrew Carnegie*, 205.
59. Standiford, *Meet You in Hell*, 57.
60. Dietrich, "Andrew Carnegie," 120.
61. Krass, *Carnegie*, 228.
62. Hessen, *Steel Titan*, 27.
63. Krass, *Carnegie*, 273.
64. Warren, *Triumphant Capitalism*, 60.
65. Schreiner, *Henry Clay Frick*, 164.
66. Deagen, "Pittsburgh Biography," 1.
67. Warren, *Industrial Genius*, 157.
68. Ibid., 14.
69. Livesay, *Andrew Carnegie*, 149.

Chapter 3

70. Farquhar, *First Million Is the Hardest*, 100.
71. Hendrick and Henderson, *Louise Whitfield Carnegie*, 50.
72. Ibid., 86.
73. Krass, *Carnegie*, 266.
74. Ibid., 243.
75. Bridge, *Carnegie Millions*, 114.
76. Hessen, *Steel Titan*, 46.
77. Casson, *Romance of Steel*, 129.
78. Wall, *Andrew Carnegie*, 78.
79. Ibid., 541.
80. Casson, *Romance of Steel*, 140.
81. Mellon, *The Judge*,ß 276.
82. Sanger, *Henry Clay Frick*, 31, 33.
83. Dietrich, "Henry Clay Frick," 81.
84. Koskoff, *Mellons*, 7.
85. Ibid., 16.
86. O'Connor, *Mellon's Millions*, 18.
87. Sanger, *Henry Clay Frick*, 47.
88. Warren, *Triumphant Capitalism*, 34.
89. Harvey, *Henry Clay Frick*, 51.
90. Ibid., 68.

91. Sanger, *Henry Clay Frick*, 79.
92. Dietrich, "Henry Clay Frick," 93.
93. Knox, *Republican Party*, 275.
94. Ibid., 276.
95. Nasaw, *Andrew Carnegie*, 310.
96. Burgoyne, *All Sorts of Pittsburghers*, "Ben Franklin Jones."
97. O'Connor, *Johnstown*, 232.
98. Sanger, *Henry Clay Frick*, 114.
99. Ingram, *Making Iron and Steel*, 72.
100. Serrin, *Homestead*, 50.
101. Sanger, *Henry Clay Frick*, 123.
102. Ibid., 132.
103. Warren, *Triumphant Capitalism*, 92.
104. Sanger, *Helen Clay Frick*, 15.
105. Ingram, *Making Iron and Steel*, 93.
106. Warren, *Triumphant Capitalism*, 88.
107. Krause, *Homestead*, 294.
108. *Commercial Gazette* archives, June 18, 1892.
109. Serrin, *Homestead*, 71.
110. Krause, *Homestead*, 19.
111. Schreiner, *Henry Clay Frick*, 80.
112. Krause, *Homestead*, 22.
113. Wall, *Andrew Carnegie*, 603.
114. Krause, *Homestead*, 31.
115. Ibid., 40.
116. Serrin, *Homestead*, 82.
117. Sanger, *Henry Clay Frick*, 187.
118. Stowell, *Fort Frick*, 185–86.
119. Josephson, *Robber Barons*, 371.
120. Standiford, *Meet You in Hell*, 209.
121. Sanger, *Henry Clay Frick*, 193.
122. Serrin, *Homestead*, 88.
123. Wall, *Andrew Carnegie*, 562.
124. Sanger, *Henry Clay Frick*, 193–94.
125. Ibid., 197.
126. Stowell, *Fort Frick*, 216.
127. Sanger, *Henry Clay Frick*, 206.
128. Stowell, *Fort Frick*, 245.
129. Vance, "New Castle Steel Story."

130. Stowell, *Fort Frick*, 286.
131. Hessen, *Steel Titan*, 39.
132. Warren, *Triumphant Capitalism*, 29.
133. Wall, *Andrew Carnegie*, 627.
134. Josephson, *Robber Barons*, 372.
135. Hessen, *Steel Titan*, 57.

Chapter 5

136. Schreiner, *Henry Clay Frick*, 210
137. Bridge, *Inside History*, 269.
138. Ingram, "Elite and Upper Class," 396.
139. Evans, *Iron Pioneer Henry W. Oliver*, 209.
140. Schreiner, *Henry Clay Frick*, 132.
141. University of Pittsburgh Carnegie Archives, December 20, 1894 correspondence.
142. Ibid., January 1, 1895 correspondence.
143. Misa, *Nation of Steel*, 83.
144. Hessen, *Steel Titan*, 65.
145. University of Pittsburgh Carnegie Archives, letter to Henry Clay Frick, February 24, 1899.
146. Ibid., Henry Clay Frick correspondence, February 24, 1899.
147. Bridge, *Inside History*, 319.
148. Hessen, *Steel Titan*, 77.
149. Serrin, *Homestead*, 154.
150. Krass, *Carnegie*, 380.
151. University of Pittsburgh Archives, Andrew Carnegie correspondence, October 1899.
152. Schreiner, *Henry Clay Frick*, 159.
153. Wall, *Andrew Carnegie*, 740.
154. Livesay, *Andrew Carnegie*, 178.
155. University of Pittsburgh Archives, letter from Andrew Carnegie to Henry Clay Frick, November 2, 1899.
156. Wall, *Andrew Carnegie*, 742.
157. Bridge, *Inside History*, 326.
158. Schreiner, *Henry Clay Frick*, 159.
159. Wheeler, *Pierpont Morgan*, 222.
160. Bridge, *Inside History*, 345.

161. Ibid., 365.
162. *Bulletin* 40, no. 17 (February 17, 1900): 11.
163. Schreiner, *Henry Clay Frick*, 173.
164. University of Pittsburgh Archives, June 10, 1898.
165. Schreiner, *Henry Clay Frick*, 174.
166. Wall, *Andrew Carnegie*, 766.
167. Hessen, *Steel Titan*, 77.

Chapter 6

168. Krass, *Carnegie*, 366.
169. University of Pittsburgh Archives, February 10, 1899.
170. Hessen, *Steel Titan*, 112.
171. Warren, *Triumphant Capitalism*, 75.
172. Wall, *Andrew Carnegie*, 767.
173. Lorant, *Pittsburgh*, 180.
174. Chernow, *House of Morgan*, 83.
175. Wall, *Andrew Carnegie*, 115.
176. Casson, *Romance of Steel*, 192.
177. Krass, *Carnegie*, 243.
178. Wheeler, *Pierpont Morgan*, 227.
179. Hessen, *Steel Titan*, 163.
180. Lorant, *Pittsburgh*, 180.
181. Schreiner, *Henry Clay Frick*, 190.
182. Warren, *Industrial Genius*, 91.
183. Wall, *Andrew Carnegie*, 791.
184. University of Pittsburgh Archives, Charles Schwab correspondence, January 5, 1889.
185. Evans, *Iron Pioneer Henry W. Oliver*, 272.
186. Ingram, "Elite and Upper Class," 409.
187. Hessen, *Steel Titan*, 128.
188. Ibid., 130.
189. Krass, *Carnegie*, 432.
190. Hessen, *Steel Titan*, 118.
191. Ibid., 119.
192. Ibid., 134–35.
193. Wall, *Andrew Carnegie*, 802.
194. Chernow, *House of Morgan*, 54.

195. Wall, *Andrew Carnegie*, 802.
196. Skrabec, *George Westinghouse*, 230.
197. Warren, *Industrial Genius*, 302.

Epilogue

198. Hessen, *Steel Titan*, 154.
199. Ibid., 153, 154.
200. Standiford, *Meet You in Hell*, 15.
201. Ingram, *Making Iron and Steel*, 178.
202. Sanger, *Henry Clay Frick*, 408.
203. Warren, *Industrial Genius*, 163.
204. Smith, "Heir Who Turned on the House of Phipps," 172.
205. White, "Phipps Gifts."
206. Sanger, *Helen Clay Frick*, 287.
207. Hessen, *Steel Titan*, 177.
208. Ibid., 228.
209. Ibid., 238–39.
210. Skrabec, *Boys of Braddock*, 214.
211. Hessen, *Steel Titan*, 298.
212. Ibid.
213. Ibid., xv.
214. Warren, *Industrial Genius*, 229–30.

BIBLIOGRAPHY

Alberts, Robert C. *The Good Provider: H.J. Heinz and His 57 Varieties*. Boston: Houghton Mifflin Company, 1973.

———. *Pitt: The Story of the University of Pittsburgh, 1787–1987*. Pittsburgh, PA: University of Pittsburgh Press, 1986.

Alef, Daniel. *Andrew Carnegie: Prince of Steel*. Kindle Edition. Titans of Fortune Publishing, 2008.

———. *Charles M. Schwab: King of Bethlehem*. Kindle Edition. Titans of Fortune Publishing, 2008.

———. *Henry Heinz: 57 Varieties and More*. Kindle Edition. Titans of Fortune Publishing, 2009.

Boegner, Peggie Phipps, and Richard Gachot. *Halcyon Days: An American Family through Three Generations*. New York: Old Westbury Gardens and Abrams, 1986.

Bridge, James H. *The Carnegie Millions and the Men Who Made Them*. London: Limpus, Baker and Company, 1903.

———. *The Inside History of the Carnegie Steel Company*. New York: Aldine Book Company, 1903.

Brignano, Mary, and J. Tomlinson Fort. *Reed Smith: A Law Firm Celebrates 100 Years*. Pittsburgh, PA: Reed Smith LLP, 2002.

Brown, Eliza Smith. *Pittsburgh Legends and Visions*. Carlsbad, CA: Heritage Media Corporation, 2002.

Brown, Mark, Lu Donnelly and David Wilkins. *The History of the Duquesne Club*. Pittsburgh, PA: Duquesne Club, 1989.

Bibliography

Burgoyne, Arthur Gordon. *All Sorts of Pittsburghers*. Pittsburgh, PA: Leader All Sorts Company, 1892.

Burnley, James. *Millionaires and Kings of Enterprise*. Philadelphia: J.B. Lippincott, 1901.

Cannadine, David. *Mellon: An American Life*. New York: Alfred A. Knopf, 2006.

Carnegie, Andrew. *The Autobiography of Andrew Carnegie*. New York: Signet Classics, 2006.

———. *The "Gospel of Wealth" Essays and Other Writings*. New York: Penguin Books, 2006.

———. *Round the World*. New York: self-published, 1879.

Carr, Charles C. *An American Enterprise*. New York: Rinehart and Company, Inc., 1952.

Casson, Herbert N. *The Romance of Steel: The Story of a Thousand Millionaires*. Freeport, NY: Books for Libraries Press, 1907.

Chernow, Ron. *The House of Morgan*. New York: Atlantic Monthly Press, 1990.

———. *Titan: The Life of John D. Rockefeller, Sr*. New York: Vintage Books, 2004.

Childs, Walter C., II. *The Life and Times of John Pitcairn*. Bryn Athyn, PA: Academy of the New Church, John Pitcairn Archives, 1999.

Corey, James B. *Memoir and Personal Recollection of J.B. Corey*. Pittsburgh, PA: Pittsburgh Printing Company, 1914.

Cotter, Arundel. *United States Steel: A Corporation with a Heart*. Garden City, NY: Doubleday Page and Company, 1921.

Covey, Stephen R. *7 Habits of Highly Effective People*. New York: Simon and Schuster, 1989.

Davis, Richard R. *The Phipps Family and the Bessemer Companies*. Nashville, TN: Turner Publishing Company, 2007.

Dennis, Stephen Neal. *Keep Tryst: The Walkers of Pittsburgh and the Sewickley Valley*. Rensselaer, NY: Hamilton Printing Company, 2004.

Denton, Frank. *The Mellons of Pittsburgh*. New York: Newcomen Society of England, 1948.

Dickson, William B. *History of Carnegie Veteran Association*. Montclair, NJ: Mountain Press, 1938.

Dienstag, Eleanor Foa. *In Good Company*. New York: Warner Books, 1994.

Dietrich, William S., II. *Eminent Pittsburghers: Profiles of the City's Founding Industrialists*. Plymouth, UK: Taylor Trade Publishing, 2011.

Dolkart, Andrew S. *Cooper-Hewitt National Design Museum: The Andrew and Louise Carnegie Mansion*. New York: Scala Publishers, Ltd., 2002.

Elkus, Leonore R., ed. *Famous Men and Women of Pittsburgh*. Pittsburgh, PA: Pittsburgh History and Landmarks Foundation, 1981.

Evans, Henry Oliver. *Iron Pioneer Henry W. Oliver, 1840–1904*. New York: E.P. Dutton and Company, Inc., 1942.

Farquhar, A.B. *The First Million Is the Hardest: An Autobiography*. Garden City, NJ: Doubleday, 1922.

Fox, Arthur B. *Pittsburgh during the American Civil War, 1860–1865*. Chicora, PA: Mechling Bookbindery, 2002.

Fox, John L. *Housing for the Working Classes: Henry Phipps from the Carnegie Steel Company to Phipps Houses*. Larchmont, NY: Memory's Tone Publishing, Inc., 2007.

Frick, Helen Clay, as told to Mary O'Hara. *My Father, Henry Clay Frick*. The Frick Art and Historical Center. Based on a series of articles in the *Pittsburgh Press*, August 1959.

Gladish, Richard R. *John Pitcairn: Uncommon Entrepreneur*. Bryn Athyn, PA: Academy of the New Church, 1989.

Gladwell, Malcolm. *Outliers*. New York: Little, Brown and Company, 2008.

Grace, Eugene Gifford. "First Annual Address: Charles M. Schwab Memorial Lectureship of the American Iron and Steel Institute." Hotel Pierre, New York, May 21, 1947.

Harper, Frank C. *Pittsburgh: Forge of the Universe*. New York: Comet Press Books, 1957.

Harvey, George. *Henry Clay Frick. The Man*. Frick Collection. New York: privately printed, 1928.

Hendrick, Burton J., and Daniel Henderson. *Louise Whitfield Carnegie*. New York: Hastings House, 1950.

Herman, Arthur. *How the Scots Invented the Modern World*. New York: Three Rivers Press, 2001.

Hessen, Robert. *Steel Titan: The Life of Charles M. Schwab*. Pittsburgh, PA: University of Pittsburgh Press, 1975.

History of Allegheny County, Pennsylvania. Vol. 11. Chicago: A. Warner and Company, 1889.

Hoerr, John P. *And the Wolf Finally Came: The Decline of the American Steel Industry*. Pittsburgh, PA: University of Pittsburgh Press, 1988.

Ingham, John N. *American National Biography*. American Council of Learners Societies. Oxford, UK: Oxford University Press, 2000.

Ingram, John N. "Elite and Upper Class in the Iron and Steel Industry, 1874–1964." Doctoral dissertation, University of Pittsburgh, 1973.

———. *The Iron Barons*. Westport, CT: Greenwood Press, 1978.

———. *Making Iron and Steel: Independent Mills in Pittsburgh, 1820–1920*. Columbus: Ohio State University Press, 1991.

Jarow, Gail. *Robert H. Jackson: New Deal Lawyer, Supreme Court Justice, Nuremberg Prosecutor*. Honesdale, PA: Calkins Creek, 2008.

Jonnes, Jill. *Empires of Light: Edison, Tesla, Westinghouse and the Race to Electrify the World*. New York: Random House, 2003.

Jordan, John W. *Encyclopedia of Biography of Pennsylvania*. New York: Lewis Historical Publishing Company, 1914.

Josephson, Matthew. *The Robber Barons*. New York: Harvest Book, 1934.

Knowles, Ruth Sheldon. *The Greatest Gamblers: The Epic of American Oil Exploration*. New York: McGraw-Hill Book Company, 1959.

Knox, Thomas Wallace. *The Republican Party and Its Leaders*. New York: P.F. Collier, 1892.

Koskoff, David E. *The Mellons: The Chronicle of America's Richest Family*. New York: Thomas Y. Crowell Publishers, 1978.

Krass, Peter. *Carnegie*. Hoboken, NJ: Peter Wiley & Sons, 2002.

Krause, Paul. *Homestead, 1880–1892*. Pittsburgh, PA: University of Pittsburgh Press, 1992.

Lamont-Brown, Raymond. *The Richest Man in the World: Carnegie*. Phoenix Mill, UK: Sutton Publishing, 2006.

Lentz, Steve. *One Life Can Make a Difference: It Was Never About the Ketchup! The Life and Leadership Secrets of H.J. Heinz*. Garden City, NY: Morgan James, 2006.

Leupp, Francis E. *George Westinghouse: His Life and Achievements*. Boston: Little, Brown, and Company, 1919.

Levine, I.E. *Inventive Wizard: George Westinghouse*. New York: Jullian Messner, 1962.

Linkon, Sherry Lee, and John Russo. *Steeltown, U.S.A*. Lawrence: University Press of Kansas, 2002.

Livesay, Harold C. *Andrew Carnegie and the Rise of Big Business*. Glenview, IL: Scott, Foresman and Company, 1975.

Lloyd, Thomas E. *History of the Jones & Laughlin Steel Corporation*. Jones and Laughlin Papers at the Heinz Museum. December 1, 1938.

Lorant, Stefan. *Pittsburgh: The Story of an American City*. Garden City, NY: Doubleday and Company Inc., 1964.

Love, Philip H. *Andrew W. Mellon: The Man and His Work*. Baltimore, MD: F. Heath Coggins and Company, 1929.

Lubove, Roy, ed. *Pittsburgh*. New York: New Viewpoints, 1976.

Lundberg, Ferdinand. *The Rich and the Super Rich*. New York: Lyle Stuart, Inc., 1968.

Mallison, Sam. *The Great Wildcatter*. Charleston: Education Foundation of West Virginia, Inc., 1953.

Malone, Dumas. ed. *Dictionary of American Biography*. Vol. 5. New York: Charles Scribner's Sons, 1904.

McCafferty, E.D. *Henry J. Heinz*. New York: Bartlett Orr Press, 1923.

McCullough, David. *The Johnstown Flood*. New York: Simon and Schuster, 1968.

McHugh, Jeanne. *Alexander Holley and the Makers of Steel*. Baltimore, MD: Johns Hopkins Press, 1980.

Meese, Hugh P. "Edgar Thomson Steel Works." In *Unwritten History: Braddock's Field*, edited by George H. Lamb. Written for the Golden Jubilee of Braddock, 1917.

Mellon, James. *The Judge: A Life of Thomas Mellon, Founder of a Fortune*. New Haven, CT: Yale University Press, 2011.

Mellon, Paul. *Reflections in a Silver Spoon*. New York: William Morrow and Company, Inc., 1992.

Mellon, Rachel. *And Allied Families*. Philadelphia: J.B. Lippincott Company, 1943.

Mellon, Thomas. *Thomas Mellon and His Times*. Pittsburgh, PA: University of Pittsburgh Press, 1994.

Mellon, William Larimer. *Judge Mellon's Sons*. N.p.: privately printed, 1948.

Misa, Thomas J. *A Nation of Steel: The Making of Modern America, 1865–1925*. Baltimore, MD: Johns Hopkins University Press, 1995.

Morell, Admiral Ben. "Jones and Laughlin: The Growth of an American Business, 1853–1953." Address at the Duquesne Club, New York, May 14, 1953.

Morell, Parker. *Diamond Jim: The Life and Times of James Buchanan Brady*. New York: AMS Press, Simon and Schuster, Inc., 1934.

Nasaw, David. *Andrew Carnegie*. New York: Penguin Press, 2006.

O'Connor, Harvey. *Mellon's Millions: The Biography of a Fortune*. New York: John Day Company, 1933.

O'Connor, Richard. *Johnstown: The Day the Dam Broke*. Philadelphia: J.B. Lippincott Company, 1957.

Prout, Henry G. *A Life of George Westinghouse*. New York: Cosimo Classics, 2005.

Rea, Henry Oliver. *Henry William Oliver, 1807–1888: Ancestry and Descendants*. Dugannon, IE: Tyrone Printing Company, 1959.

Reed, George Irvine, ed. *Century Cyclopedea of History and Biography in Pennsylvania*. Vol. 2. Chicago: Century Publishing and Engraving Company, 1904.

Rottenberg, Dan. *In the Kingdom of Coal*. New York: Routledge, 2003.

Samuel, Larry. *Rich: The Rise and Fall of American Wealth*. New York: Amacon, 2009.

Sanger, Martha Frick Symington. *Helen Clay Frick: Bittersweet Heiress*. Pittsburgh, PA: University of Pittsburgh Press, 2007.

———. *Henry Clay Frick: An Intimate Portrait*. New York: Abbeville Press Publishers, 1998.

Schreiner, Samuel. *Henry Clay Frick: The Gospel of Greed*. New York: St. Martin's Press, 1995.

Schribman, David, and Angelika Kane, eds. *Pittsburgh Lives*. Chicago: Triumph Books, 2006.

Schwab, Charles M. *Succeeding with What You Have*. New York: Cosimo Books, 2005.

Serrin, William. *Homestead: The Glory and Tragedy of an American Steel Town*. New York: Times Books, 1992.

Skrabec, Quentin R., Jr. *The Boys of Braddock: Andrew Carnegie and the Men Who Changed Industrial History*. Westminster, MD: Heritage Books, 2004.

———. *George Westinghouse: Gentle Genius*. New York: Algora Publishing, 2007.

———. *H.J. Heinz*. Jefferson, NC: McFarland and Company, 2009.

———. *The Metallurgic Age: The Victorian Flowering of Invention and Industrial Science*. Jefferson, NC: McFarland and Company, 2006.

———. *The World's Richest Neighborhood: How Pittsburgh's East Enders Forged Industry*. New York: Algora Publishing, 2010.

Standiford, Les. *Meet You in Hell: Andrew Carnegie, Henry Clay Frick and the Bitter Partnership That Transformed America*. New York: Crown Publishers, 2005.

Stiles, T.J. *The First Tycoon: The Epic Life of Cornelius Vanderbilt*. New York: Alfred A. Knopf, 2009.

Stowell, Myron R. *Fort Frick: The Siege of Homestead*. Pittsburgh, PA: Pittsburgh Printing Company, 1893.

Stubbles, John R. *The Original Steelmakers*. Warrendale, PA: Iron and Steel Society, 1984.

Swetnam, George, and Helene Smith. *The Carnegie Nobody Knows*. Greensburg, PA: McDonald/Sward Publishing, 1989.

Tarbell, Ida. *The Life of Elbert H. Gary: The Story of Steel*. New York: D. Appleton and Company, 1925.

Tedlow, Richard S. *Giants of Enterprise: Seven Business Innovators and the Empires They Built*. New York: Harper Business Books, 2001.

Temin, Peter. *Iron and Steel in Nineteenth-Century America*. Cambridge, MA: MIT Press, 1964.

Vukmir, Dr. Rade B. *The Mill*. Lanham, MD: University Press of America Inc., 1999.

Wall, Joseph Frazier. *Andrew Carnegie*. Pittsburgh, PA: University of Pittsburgh Press, 1989.

Ward, James A. *J. Edgar Thomson: Master of Pennsylvania*. Westport, CT: Greenwood Press, 1980.

Warren, Arthur. *George Westinghouse, 1846–1914*. N.p.: Westinghouse Publicity Department, 1923.

Warren, Kenneth. *Industrial Genius: The Working Life of Charles Michael Schwab*. Pittsburgh, PA: University of Pittsburgh Press, 2007.

———. *Triumphant Capitalism: Henry Clay Frick and the Industrial Transformation of America*. Pittsburgh, PA: University of Pittsburgh Press, 1996.

———. *Wealth, Waste and Alienation: Growth and Decline in the Connellsville Coke Industry*. Pittsburgh, PA: University of Pittsburgh Press, 2001.

Wheeler, George. *Pierpont Morgan and Friends: The Anatomy of a Myth*. Englewood Cliffs, NJ: Prentice Hall, Inc., 1973.

Wilhelm, Stephen R. *Men of Petroleum Progress*. Houston: Golden Anniversay Publication, 1952.

Wilmerding World Wide. *Wilmerding and the Westinghouse Air Brake Company*. Charleston, SC: Arcadia Publishing, 2002.

Wilson, Erasmus, ed. *Standard History of Pittsburgh*. Pittsburgh, PA: H.R. Cornell and Company, 1898.

Wollman, David H., and Donald R. Inman. *Portraits in Steel*. Kent, OH: Kent University Press, 1999.

Videos and Films

Bussler, Mark, prod. and written by. *Westinghouse*. Inecom Entertainment, 2008.

Land, Stephen. *Empires of Industry: Andrew Carnegie and the Age of Steel*. History Channel, 1997.

Process: Change and Sacrifice: Inclination and the Walk through Fire. 2003.

Redinger, David. *Pillars of Fire*. Westmoreland-Fayette Historical Society. New American Films, 2007.

Weaver, Fritz, narr. *Heinz: The Story of an American Family*. QED Communications, Inc., for WQED, 1992.

Periodicals and Internet Publications

American Experience. "People and Events: Andrew Carnegie." PBS.

Barcousky, Len. "The Gospel According to Andy." *Pittsburgh Post Gazette*, October 26, 2006.

Blatz, Perry K. "Pittsburgh: The Fiery Scapegoat for the Country." *Western Pennsylvania History* (Fall 2011).

Bridge, James H. "Captains of Industry: Henry Phipps." *Cosmopolitan*, October 1903.

Clarksburg Exponent-Telegram, July 5, 1953.

———. "Portrait of Michael Benedum." October 13, 1959.

Clarksburg Telegram, October 20, 1949.

Deagen, Brian. "Pittsburgh Biography." *Investor's Business Daily*, October 6, 2006.

Detar, James. "Westinghouse Was Plugged In." *Investor's Business Daily*, October 26, 2010.

Dietrich, William S., II. "Andrew Carnegie." *Pittsburgh Quarterly* (Summer 2007).

———. "Andrew W. Mellon." *Pittsburgh Quarterly* (Fall 2007).

———. "Henry Clay Frick." *Pittsburgh Quarterly* (Spring 2009).

———. "Michael Late Benedum: A Character Portrait in Oil." *Pittsburgh Quarterly* (Fall 2009).

———. "Money, Power & Purpose: The Story of Joseph Kennedy." *Pittsburgh Quarterly* (Spring 2011).

———. "The Mystery of George Westinghouse." *Pittsburgh Quarterly* (Summer 2006).

———. "Smiling Charlie Schwab." *Pittsburgh Quarterly* (Spring 2010).

Donnelly, Lu. "Pittsburgh Bathhouses." *Western Pennsylvania History Magazine*, Winter 2011–12.

Felten, Eric. "Now College Is the Break." *Wall Street Journal*, February 11, 2011.

Fitzpatrick, Dan. "Mellon Family Member: Patriarch Would Have Lamented Bank Deal." *Pittsburgh Post-Gazette*, October 18, 2007.

———. "The Voice of Carnegie." *Pittsburgh Post Gazette*, October 30, 2007.

Garlin, Hamlin. "Homestead and Its Perilous Trades." *McClure's Magazine*, June 1894.

Giddens, P.H. "Men Who Have Made Oil History." Pennsylvania Historical and Museum. Drake Well Memorial Park, PA, 1944.

Goebel, Greg. "George Westinghouse." Wikipedia. June 6, 2006.

Graham, Jed. "Carnegie Was Tough as Steel." *Investor's Business Daily*, June

2, 1910.

Griffith, William. "Henry Phipps: The Man and His Millions." *New York Times* magazine section, April 2, 1905.

Imerito, Tom. "The Westinghouse Legacy." *Pittsburgh TEQ* 13, no. 9 (November 2007).

Iron Age 21 (May 1903).

Jones, Ben Franklin. "Two Messages." *North American Review* 146, no. 374 (January 1888).

McGuire, Patrick. "Smitten by St. Louis's Other Arches." *Wall Street Journal*, February 5, 2011.

The Pioneer 15, no. 8. "Portrait of Michael Late Benedum." (1959).

Pittsburgh Bulletin, 1900–5.

Pittsburgh Post, May 21, 1903.

Pitz, Marylynne. "The Domestics." *Pittsburgh Post Gazette*, March 23, 2011.

Randolph, Jennings. "I Kept Right on Going." Reprint from Success Unlimited, 1955.

Schutte, Laureen, ed. *Bulletin*. Pittsburgh History & Landmarks Foundation. James D. Van Trump Library, 1887–1900.

Smith, Richard Austin. "The Heir Who Turned on the House of Phipps." *Fortune*, October 1960.

Tannler, Albert M. "Renaissance Man: Architect Grosvenor Atterbury Designed Showcases, Housing for the Working Poor." *Focus*, April 11, 2004.

Tucker, Abegail. "Cutthroat Capaitalist." *Smithsonian*, January 2011.

Vance, Clifford. "The New Castle Steel Story." Supplement to the *New Castle News*, 1943.

Westinghouse, George, IV. "The Important Breakthrough That Never Happened." *News & Views* 2006–5.

White, William A. "Phipps Gifts." *Pittsburgh Press*, January 26, 1956.

Zubrinic, Darko. "Joe Magarac." Crown Croatian World Network. www.Coatia.org. March 18, 2008.

Other

B.F. Jones Memorial Library Archives of Aliquippa, Pennsylvania.

Bridgeport Public Library Benedum Collection. Bridgeport, West Virginia.

Frick Art and Historical Center Archives. "A Gallery Guide to the Exhibition from J.P. Morgan to Henry Clay Frick." Pittsburgh, Pennsylvania, 2007.

Glencairn Museum Archives. Academy of the New Church. Bryn Athyn,

Pennsylvania. Dirk Junge and Gregory Jackson (archivist).
Mancino, Dr. Pete, graphologist.
National McKinley Birthplace Memorial and Library. Niles, Ohio.
Sewickley Heights Museum, W.L. Jones Papers and B.F. Jones Files.
Sewickley Valley Historical Society.
Steel Industry Heritage Corporation, Bost Building. Homestead, Pennsylvania.
University of Pittsburgh Archives Service Center. Helen Clay Frick Foundation Archives. Pittsburgh, Pennsylvania.

ABOUT THE AUTHOR

Dale Richard Perelman has written several books, including *Mountain of Light*, *The Regent* and *Centenarians*. Mr. Perelman holds a bachelor of arts degree in English literature from Brown University and an MBA in industrial relations from the Wharton School of the University of Pennsylvania.